WORKING WITH PLASTICS

Other Publications:

PLANET EARTH
COLLECTOR'S LIBRARY OF THE CIVIL WAR
LIBRARY OF HEALTH
CLASSICS OF THE OLD WEST
THE EPIC OF FLIGHT
THE GOOD COOK
THE SEAFARERS
THE ENCYCLOPEDIA OF COLLECTIBLES
THE GREAT CITIES
WORLD WAR II
THE WORLD'S WILD PLACES
THE TIME-LIFE LIBRARY OF BOATING
HUMAN BEHAVIOR
THE ART OF SEWING
THE OLD WEST
THE EMERGENCE OF MAN
THE AMERICAN WILDERNESS
THE TIME-LIFE ENCYCLOPEDIA OF GARDENING
LIFE LIBRARY OF PHOTOGRAPHY
THIS FABULOUS CENTURY
FOODS OF THE WORLD
TIME-LIFE LIBRARY OF AMERICA
TIME-LIFE LIBRARY OF ART
GREAT AGES OF MAN
LIFE SCIENCE LIBRARY
THE LIFE HISTORY OF THE UNITED STATES
TIME READING PROGRAM
LIFE NATURE LIBRARY
LIFE WORLD LIBRARY
FAMILY LIBRARY:
   HOW THINGS WORK IN YOUR HOME
   THE TIME-LIFE BOOK OF THE FAMILY CAR
   THE TIME-LIFE FAMILY LEGAL GUIDE
   THE TIME-LIFE BOOK OF FAMILY FINANCE

This volume is part of a series offering homeowners
detailed instructions on repairs, construction
and improvements they can undertake themselves.

HOME REPAIR
AND IMPROVEMENT

WORKING WITH PLASTICS

BY THE EDITORS OF
TIME-LIFE BOOKS

TIME-LIFE BOOKS
ALEXANDRIA, VIRGINIA

Time-Life Books Inc.
is a wholly owned subsidiary of
TIME INCORPORATED

Founder	Henry R. Luce 1898-1967

Editor-in-Chief	Henry Anatole Grunwald
President	J. Richard Munro
Chairman of the Board	Ralph P. Davidson
Executive Vice President	Clifford J. Grum
Chairman, Executive Committee	James R. Shepley
Editorial Director	Ralph Graves
Group Vice President, Books	Joan D. Manley
Vice Chairman	Arthur Temple

TIME-LIFE BOOKS INC.

Managing Editor	Jerry Korn
Text Director	George Constable
Board of Editors	Dale M. Brown, George G. Daniels, Thomas H. Flaherty Jr., Martin Mann, Philip W. Payne, Gerry Schremp, Gerald Simons
Planning Director	Edward Brash
Art Director	Tom Suzuki
Assistant	Arnold C. Holeywell
Director of Administration	David L. Harrison
Director of Operations	Gennaro C. Esposito
Director of Research	Carolyn L. Sackett
Assistant	Phyllis K. Wise
Director of Photography	Dolores A. Littles

Chairman	John D. McSweeney
President	Carl G. Jaeger
Executive Vice Presidents	John Steven Maxwell, David J. Walsh
Vice Presidents	George Artandi, Stephen L. Bair, Peter G. Barnes, Nicholas Benton, John L. Canova, Beatrice T. Dobie, Carol Flaumenhaft, James L. Mercer, Herbert Sorkin, Paul R. Stewart

HOME REPAIR AND IMPROVEMENT

Editor	Robert M. Jones
Senior Editor	Betsy Frankel
Designer	Ed Frank

Editorial Staff for Working with Plastics

Text Editors	Victoria W. Monks, Brooke Stoddard (principals), Rachel Cox, Peter Pocock, Mark M. Steele
Writers	Tim Appenzeller, Kevin Armstrong, Carol Jane Corner, Stuart Gannes, Kathy Kiely, Kirk Y. Saunders
Copy Coordinator	Diane Ullius Jarrett
Art Assistants	George Bell, Fred Holz, Lorraine D. Rivard, Peter Simmons
Editorial Assistant	Cathy A. Sharpe

Editorial Operations

Production Director	Feliciano Madrid
Assistants	Peter A. Inchauteguiz, Karen A. Meyerson
Copy Processing	Gordon E. Buck
Quality Control Director	Robert L. Young
Assistant	James J. Cox
Associates	Daniel J. McSweeney, Michael G. Wight
Art Coordinator	Anne B. Landry
Copy Room Director	Susan B. Galloway
Assistants	Celia Beattie, Ricki Tarlow

Correspondents: Elisabeth Kraemer (Bonn); Margot Hapgood, Dorothy Bacon (London); Susan Jonas, Lucy T. Voulgaris (New York); Maria Vincenza Aloisi, Josephine du Brusle (Paris); Ann Natanson (Rome).

THE CONSULTANTS: Karl Plitt is retired from the National Bureau of Standards, where he was a physical science administrator serving as chief of the plastics and textiles sections. He has received many awards for his work in plastics technology, owns plastics patents, and is the author of numerous publications in this field.

Paul Albert is a fiberglass sculptor in Washington, D.C. He has had sculptures on display at the Hirshhorn Museum in Washington and teaches at Northern Virginia Community College.

Roswell W. Ard is a consulting structural engineer and a professional home inspector in northern Michigan. He has written professionally on construction techniques.

Harold Campbell is a designer and producer of display cases for seven museums of the Smithsonian Institution in Washington, D.C. He has worked in plastics designing and production since 1973.

Paul Campbell is retired from the National Bureau of Standards in Washington, D.C., where he was a supervisory research chemist dealing with paints and coatings. He has written extensively on this subject.

Harris Mitchell, special consultant for Canada, has worked in the field of home repair and improvement since 1950. He is Homes editor of *Today* magazine, writes a syndicated newspaper column, "You Wanted to Know," and is the author of a number of books on home improvement.

For information about any Time-Life book, please write:
Reader Information
Time-Life Books
541 North Fairbanks Court
Chicago, Illinois 60611

Library of Congress Cataloguing in Publication Data
Main entry under title:
Working with plastics.
   (Home repair and improvement; 31)
   Includes index.
   1. Plastics craft. I. Time-Life Books.
II. Series.
TT297.W67   691'.92   81-18459
ISBN 0-8094-3506-3   AACR2
ISBN 0-8094-3507-1 (lib. bdg.)
ISBN 0-8094-3509-8 (pbk.)

Contents

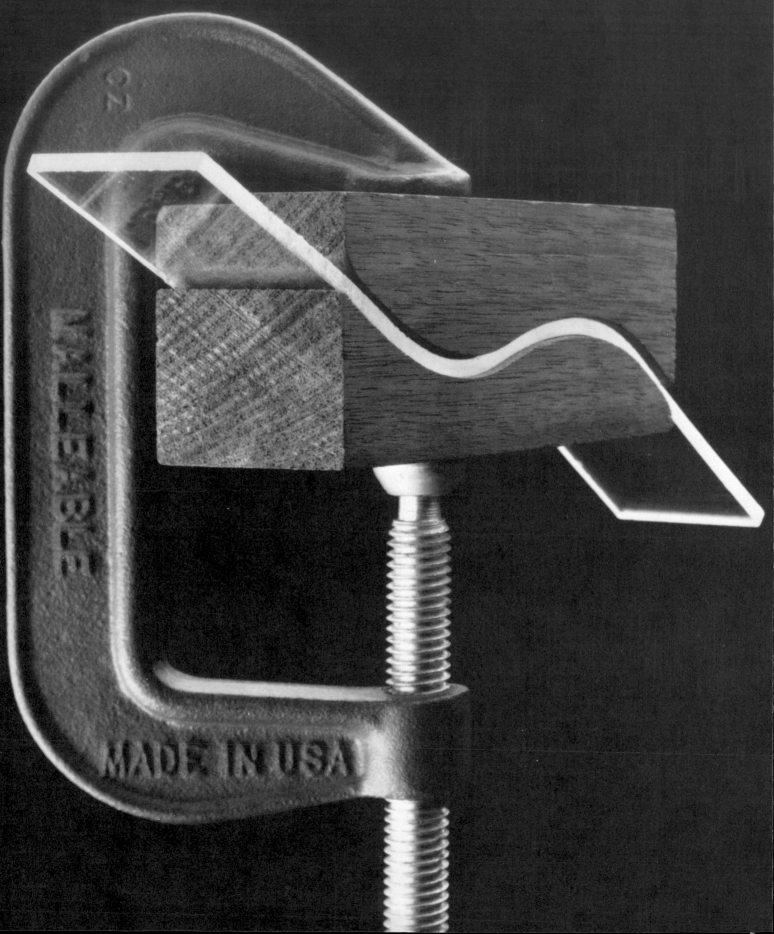

Squeeze play. An oven-softened strip of sheet acrylic is molded into an S shape by means of a wood block sawed into two matching parts. The C clamp will hold the two parts of the jig together until the plastic cools and hardens—a matter of minutes. All thermoplastics, of which acrylic is one, have this capacity to be shaped by a combination of gentle heat and pressure.

In 1863, a New York printer named John Wesley Hyatt set out to concoct a substitute for ivory, hoping to win a $10,000 prize offered by the American billiard-ball maker Phelan & Collender. Hyatt's tinkerings yielded celluloid—too brittle, alas, to displace elephant tusks as the source of billiard balls, but perfect for ladies' combs and gentlemen's collars. Celluloid was the first of the plastics, a family of materials that now touch every aspect of our lives.

Once regarded as cheap substitutes for "real materials," plastics today do a tremendous assortment of jobs better than any natural substance can. The layers in a sheet of plywood are bonded by plastic resins, the strongest of all adhesives. Gossamer-thin sheets of polyester film cover windows to form solar screens that block out 80 per cent of the sun's rays without interfering with the view. Tough, durable nylon has replaced metal in gears for many home appliances because the smooth, waxy texture of nylon gears allows them to turn quietly and efficiently without a drop of oil. Lightweight foamed plastics, sandwiched into concrete blocks and panels, serve to insulate the foundations and walls of a house.

Amazingly, almost all of these plastics and countless others are derived from one source—petroleum. Plastics are made by breaking down the components of crude oil into simple molecules made up of hydrogen and carbon atoms, then stringing these units into long molecular chains *(page 97)*.

The unique and widely variable forms that plastics take, as well as their lightness, strength and malleability, result in part from the variety of molecules that can be derived from petroleum and in part from the different methods chemists have devised for linking them. To further increase the range of possibilities, chemists can change the characteristics of plastics with various additives. Just six years after John Hyatt created celluloid, camphor was added to that bone-hard plastic, rendering it resilient enough for use in billiard balls. Today, the addition of microscopic whiskers of boron to epoxy resins produces a material so strong that it is used to make helicopter rotor blades.

Despite their chameleon qualities and their bewildering variety of characteristics, plastics are, on the whole, quite manageable materials. Many of the familiar tools and techniques that are employed in working with wood and metal can also be used to cut, file, shape, join and mold these synthetic substances. Indeed, plastics are so adaptable to the skills and needs of the amateur that, as sheathings, adhesives, coatings and moldable resins, they lend themselves to literally hundreds of small and large projects designed to improve and repair the home.

A Beginner's Guide to Manipulated Molecules

Plastics, once scorned as cheap substitutes for natural materials, are now welcomed throughout the home for qualities that in many cases make them superior to the materials they replace. Rigid plastics serve as patio roofs, plumbing pipes, lighting fixtures and furniture. Flexible varieties go into garden hoses, upholstery, dropcloths, and food and beverage containers. As tiny particles or liquid resins, plastics are found in many paints, sealants, caulks and adhesives. Other resins can be cast or molded into hard, durable doorknobs and tool handles, or they can be bubbled into rigid foams for insulation and insulated cups, or into supple ones for mattresses and pillows.

Despite plastics' bewildering variety, each kind—as shown in the chart on page 12—offers a unique set of properties that can be used to advantage by the amateur. The most notable property is the one implied by their name: At some point in their manufacture, all plastics are truly plastic—either fully liquid or pliant enough to be formed and molded. They are organized into two distinct groups according to the circumstances under which they become plastic, or capable of being molded.

Plastics of the first group, the thermosetting plastics, are moldable only once, after which they solidify through a chemical process. These plastics, familiar as epoxy cements and as the hardened resins used to surface such laminates as Formica, cannot be reheated to moldable softness; high temperatures merely char and decompose them. Those in the second group, the thermoplastics, regain their pliancy each time they are heated and can be remolded repeatedly. They are most familiar as the vinyls used in house siding and the polyethylene of food-storage bags, as well as the hard, glossy polystyrene plastics out of which appliance casings are molded and the clear, tough acrylic and polycarbonate of shatterproof windows.

Both kinds of plastic behavior are a boon to the amateur. Thermosetting plastics often come as liquid resins—epoxy and polyester are commonly available—that are easy to cast at room temperature. Mixed with hardener, they set to form hard and heat-resistant products ranging

from a doorknob to a shower enclosure reinforced with glass fibers.

Solid thermoplastics, on the other hand, can be softened with moderate heat, then twisted, creased or bent. A seemingly rigid sheet of acrylic glass, for instance, can be softened in a kitchen oven to make fittings and furniture parts that are both decorative and practical.

Working with plastics involves many of the same shaping and finishing operations used in metalworking and woodworking. Thermoplastics parallel metals in that they can be heated for bending and can be welded. Like wood, plastics can be cut with hand and power saws, drilled, shaped with files and smoothed with sandpaper. These familiar processes, however, can sometimes take unfamiliar twists when they involve plastics. Many plastics, for example, soften at heats so moderate that the friction of a power saw, drill or sander may be enough to melt them and gum the tool. Other plastics are so flexible that they distort when sawed or drilled, making it difficult to do accurate work. The specific remedies for these problems are presented in the pages that follow.

In many instances, the unusual characteristics that present problems can also be turned to advantage. Exposed to strong cleaning solutions and solvents, for example, many plastics—including most of the thermoplastics and especially polystyrene—weaken, crack and soften. Although this means that painstaking care must be used in matching solvent-based glues and paints to the plastic surfaces, it also makes possible a unique bonding process called solvent cementing. A bead of solvent, run along the seam, softens the adjacent plastic surfaces. The edges then meld, and the bond hardens as the solvent evaporates.

Because they are fluid in some stage of their manufacture or use, plastics can be colored, combined with additives or reinforcements, or frothed into foams, all of which further broadens the range of their properties and applications. Additives that can be used at home include mineral fillers that make casting resins thicker and easier to work with, as well as coloring agents. And glass-reinforced polyester is easily fabricated at home by the layering of mats of glass fiber soaked with polyester resin.

Basic Safety Rules for Working with Plastics

Working with plastics requires many of the precautions that are standard in woodworking and metalworking. Guards on power tools and heating equipment should be checked and maintained. Safety glasses are a must, and a dust mask is essential for operations that produce powdery residues. In addition, plastics call for precautions against two special dangers: irritation from fumes and flammability.

With liquid plastics—the resins, hardeners, paints, sealants and adhesives—both hazards are immediate. While working with these liquids, ensure proper ventilation to disperse fumes. In many cases, you should wear a respirator with a charcoal filter (page 69). Do not smoke, and never work near an open flame. Solid plastics may generate harmful vapors, too, if they are allowed

to become overheated during welding or hot-bending operations.

To avoid skin contact with the liquids, wear rubber gloves and a shop apron while pouring or mixing. Proper clothing is also important for sanding, drilling and sawing reinforced plastics—especially those that include glass; tiny, razor-sharp particles are released during these operations and will quickly irritate skin, eyes and lungs. Gloves, a shop apron and a good-quality dust mask are essential.

Many solid plastics are somewhat flammable, although no solid plastic in common use today presents any greater fire hazard than wood. The fumes of burning plastics, however, can be toxic. Keep a multipurpose, dry chemical fire extinguisher nearby, especially if you are working with liquid plastics.

The Substances of Everyday Objects

Household article	Plastic
Adhesives	
Contact cements, flexible adhesives	Silicones, synthetic rubbers
Plastic cements	Cellulosics, polystyrene
Two-component cements	Epoxies
White glue	PVAC
Appliance housings	ABS, polystyrene
Bathtub and shower enclosures	Polyester, glass-reinforced (fiberglass)
Bottles, containers	
Translucent or opaque; flexible	Polyethylene, polypropylene
Clear	Polyester, PVC
Buckets, washtubs	Polyethylene, polypropylene
Countertops	
Surfacing	Phenolic/melamine laminate (Formica,
Artificial marble	Micarta) Acrylic (Corian), polyester
Cushions, pillows, mattresses	Polyurethane foam, PVC foam
Decorative clear castings	Acrylic, polyester
Dishes	Melamine (Melmac)
Dishwasher and washing-machine interiors	Polypropylene
Drinking glasses	
Clear and rigid	Polystyrene
Flexible	Polyethylene
Insulated cups	Polystyrene foam (Styrofoam)
Electrical circuit boards, switch toggles	Paper- and cotton-reinforced and laminated epoxies and phenolics
Fillers	
Caulking compounds	Polyurethane, PVAC, silicones, synthetic rubbers
Grouts	PVAC, silicones
Mortars	Epoxies, PVAC
Putties, filler pastes	Epoxies, polyester, PVAC
Patching kits	Polyester, glass-reinforced (fiberglass)
Spackling compound	PVAC
Films	
Colored artist's films	Cellulosics
Food wrap	Polyethylene, polypropylene, PVDC (Saran)
Magnetic tape	Polyester (Mylar)
Photographic film	Cellulosics
Flooring (sheet and tiles)	PVC
Furniture	
Clear	Acrylic (Lucite, Plexiglas)
Flexible	Polypropylene
Glossy and opaque	ABS; polyester, glass-reinforced (fiberglass)
Upholstered	PVC (Naugahyde)
Garden hoses	PVC
Gutters and downspouts	ABS, PVC

Plastics in the home. The left-hand column of the chart at left lists the plastic household objects most likely to confront someone engaged in home crafts. To the right are the plastics most often used for each object. In cases where a plastic is known by an abbreviation—ABS for acrylonitrile-butadiene-styrene, PVC for polyvinyl chloride and PVDC for polyvinylidene chloride are examples—the abbreviation is listed. Similarly, common names and widely recognized brand names are included in parentheses where helpful. In some cases, nearly identical objects are made of different kinds of plastic, depending on the manufacturer. When repairing such objects, you can tell similar plastics apart by consulting the chart on page 12 and matching the plastic at hand to the one on the chart whose characteristics it most closely resembles.

continued on page 10

continued from page 9

Household article	Plastic
Handles and knobs	
Clear	Acrylic, cellulosics
Black, pots and pans	Reinforced phenolics
Insulation foams	
Preformed	Polystyrene (Styrofoam)
Foamed-in-place	Polyurethane
Lighting	
Fixtures	Reinforced phenolics
Shades, diffusers	Acrylic (Lucite, Plexiglas), cellulosics, polycarbonate (Lexan, Merlon)
Lubricants	Silicones
Paints, sealants and coatings	
Ceramic and glass	Epoxies, polyurethane
Masonry	Acrylic, epoxies, PVAC, synthetic rubbers
Metal	Epoxies, phenolics
Plaster, wallboard	Acrylic, polyester, PVAC
Plastic	Polyester
Tool handles	PVC/PVAC
Wood	Acrylic, polyester, PVAC
Plumbing pipes	ABS, polyethylene, PVC
Roofing	
Patio covers	Polyester, glass-reinforced (fiberglass)
Waterproof membranes	Synthetic rubbers
Siding and paneling	PVC
Telephone housings	ABS
Toys	
Glossy and opaque	ABS, polystyrene
Flexible	Polyethylene, polypropylene
Clear and hard	Polystyrene
Tubing	Polyethylene, PVC
Wallpaper	Polyester (Mylar), PVC
Window frames	PVC
Windowpanes, skylights, clear panels	Acrylic (Lucite, Plexiglas), polycarbonate (Lexan, Merlon)

Where and How to Buy Stock

Despite the exotic names of the plastics in the chart on page 12, you can buy many of them without a trip to a specialized plastics-supply house. Glass-supply and hardware stores stock sheets of clear acrylic and polycarbonate for glazing, usually in thicknesses of ⅛ to ⅜ inch. Lumberyards carry PVC pipe, corrugated sheets of glass-reinforced polyester used for covering patios, decorative laminates for countertops and cabinet facings, and polyethylene film. Large lumberyards, hardware stores and paint stores are the best sources for many plastic-base coatings, adhesives and fillers.

Even if the type of plastic you need should call for a trip to a specialized plastics dealer, there are likely to be several listed in the classified telephone directory of any major city under "Plastics Supply Centers." Although many of them technically are wholesalers, you will usually not be required to buy in wholesale quantities.

Most plastic suppliers will cut rod, sheet and tubing, all sized in inches, to your order. But film—plastic sheets .01 inch or less in thickness and more commonly sized in mils (a mil equals .001 inch)—is available in rolls of standard widths and lengths.

In choosing a plastic, compare your re quirements with the characteristics of the plastics listed on page 12. Take into account the stresses to be placed on the finished object, including the weight it will bear, its exposure to impact, abrasion and corrosive liquids, and whether it will have to endure high temperatures and direct sunlight. Also consider the desired appearance of the plastic object.

If the characteristics you seek do not match any plastic on the chart, chances are you are looking for a laminate or for a reinforced—or filled—plastic. The laminates, which are far tougher than most pure plastics, consist of layers of nonplastic materials, such as paper or cloth, bonded together with intervening layers of plastic. Filled or reinforced plastics contain materials other than plastic—most often, powdered wood or cotton fibers—that are randomly dispersed through the plastic before it sets. Such loose, fibrous fillers impart a degree of strength and resistance to chipping that is difficult to attain in pure plastic.

A third type of hybrid, structural foam, consists of a core of foamed plastic sheathed in a solid, unfoamed skin of the same plastic. The structural foams offer a tough, attractive exterior and exceptional stiffness and may be made from many sorts of plastics.

Basic plastics offspring. Layers of coarse brown kraft paper soaked in phenolic resin form the base of a tough, moisture-proof decorative laminate (below), ideal for countertops and cabinet facings; the penultimate layer, just beneath a surface film of clear, tough melamine, is a sheet of colored or patterned paper. To make fiberglass, alternating layers of glass-fiber mat and woven glass-fiber fabric are impregnated with polyester resin. A layer of clear resin on top of the glass-reinforced layers hardens into a smooth, tough surface.

Wood or cotton fibers, scattered randomly throughout a phenolic resin, form a reinforced plastic with sufficient strength and chip-resistance to be used for coffeepot knobs and electric-switch toggles. Unlike the filler in many other reinforced plastics, this filler material does not form a distinct layer in the finished plastic.

In a typical structural foam, a smooth skin of plastic covers the multicelled core. The skin is formed during manufacture when the foaming resin, injected into a closed mold under high pressure, compacts against the mold walls, yielding a light, stiff product. Structural foam is increasingly used for prefabricated housing components, such as window frames and doorframes, shutters and shingles.

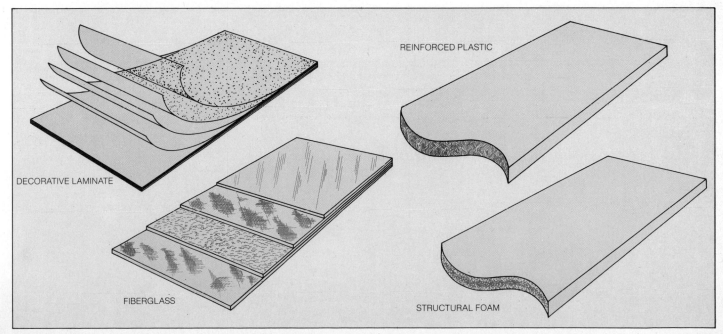

DECORATIVE LAMINATE

FIBERGLASS

REINFORCED PLASTIC

STRUCTURAL FOAM

Performance Profiles of the Major Families

Type of plastic	Characteristics	How supplied	Techniques
ABS (Acrylonitrile-butadiene-styrene)	Tough, hard, impact-resistant and resistant to chemicals, yet glossy, translucent and attractive.	Rods, sheets, pipes	Cutting, page 20 Drilling, page 30 Bending, page 32 Fastening, pages 44, 48 Finishing, page 37
Acrylic	Prized for crystal clarity and resistance to weather and impact. Scratches easily.	Rods, sheets, tubing, casting resins, in coatings	Cutting, page 20 Drilling, page 30 Bending, page 32 Fastening, pages 44, 48 Finishing, page 37 Casting, page 84 Using in coatings, page 116
Cellulosics	Flexible and tough. Transparent and easily colored. Soften at low heat.	Film, rods, sheets	Cutting, page 20 Drilling, page 30 Fastening, pages 44, 48
Epoxies	Strong; resistant to heat and chemicals, particularly when reinforced with glass fiber. As adhesives, will bond to nearly any surface. Relatively expensive.	Resins; in adhesives, coatings, fillers	Casting, page 84 Using in adhesives, coatings, fillers, pages 48, 104, 114, 116
Melamine	Valued for its transparency and extreme resistance to scratches, heat and chemicals. Generally laminated with paper or mixed with powdered wood or cotton fibers to reinforce it and reduce its brittleness.	Reinforced or laminated sheets	Cutting, page 20 Drilling, page 30 Fastening, pages 48, 98
Nylon	Tough, resistant to heat and chemicals. Slippery, self-lubricating and translucent. Much used for kitchen utensils and for gears in small appliances.	Rods, sheets	Cutting, page 20 Drilling, page 30 Fastening, pages 54, 58
Phenolics	Black or brown in color, hard, heat-resistant. To reduce their brittleness, they are reinforced with powdered wood or cotton fibers or are laminated with paper or cloth.	Reinforced or laminated rods, sheets, or tubing; in coatings	Cutting, page 20 Drilling, page 30 Fastening, pages 48, 98 Using in coatings, pages 116, 118
Polycarbonate	Similar to acrylic but tougher and more impact-resistant. Less clear, however, and more prone to scratching.	Rods, sheets, tubing	Cutting, page 20 Drilling, page 30 Bending, page 32 Fastening, pages 44, 48 Finishing, page 37
Polyester	Frequently reinforced with glass fiber for strength. Easy to cast and often used for large objects. Transparent, but may yellow in sunlight.	Film; reinforced sheets; casting resins; in coatings, fillers	Cutting, page 20 Drilling, page 30 Fastening, page 48 Casting, page 84 Reinforcing, page 90 Using in coatings, fillers, pages 104, 116 Finishing, page 37
Polyethylene	Flexible, lightweight, translucent; waxy to the touch. Softens at low heat, cracks in sunlight, and is permeable to odors and some gases.	Film, rods, sheet, pipes, tubing	Cutting, page 20 Drilling, page 30 Bending, page 32 Fastening, pages 44, 54, 58, 64 Finishing, page 37

Type of plastic	Characteristics	How supplied	Techniques
Polypropylene	Similar to polyethylene but harder and more resistant to heat.	Film, rods, sheets, tubing	Cutting, page 20 Drilling, page 30 Bending, page 32 Fastening, pages 44, 54, 58 Finishing, page 37
Polystyrene	Hard, crystal-clear and inexpensive. Brittle; makes a distinctive clink when struck. Softens at low heat; may crack in sunlight.	Rods, sheets, foam	Cutting, page 20 Drilling, page 30 Bending, page 32 Fastening, pages 44, 48, 64 Finishing, page 37
Polyurethane	Tough, flexible and resistant to chemicals in both its solid and its foamed forms. Often foamed in place as insulation, but like most plastics it is not fireproof and, in burning, gives off poisonous gases. Opaque.	Rods; sheets; foam; foaming compounds; in coatings, fillers	Cutting, page 29 Fastening, page 48 Using in coatings, fillers, pages 114, 116
PVAC (Polyvinyl acetate)	An important component in adhesives, paints and fillers, often chemically combined with PVC. Flexible; adheres readily to other materials.	In coatings, fillers, adhesives	Using in coatings, adhesives, fillers, pages 48, 114, 116
PVC (Polyvinyl chloride)	Pliant, elastic, but in combination with various additives can be either flexible or rigid. Transparent, softens at low heat.	Film, sheets, rods, tubing, pipes, in coatings	Cutting, page 20 Drilling, page 30 Fastening, pages 44, 48, 54, 58, 64 Using in coatings, page 116 Finishing, page 37
PVDC (Polyvinylidene chloride)	Remains strong even when extremely thin, and provides an excellent barrier against moisture and gases. Softens at low heat.	Film	
Silicones	As elastic compounds, tough and slippery to the touch. As a coating, resistant to heat and chemicals. Expensive.	Resins; in molding compounds, coatings, lubricants, adhesives, fillers	Using to make flexible molds, page 78 Using in coatings, page 116 Using in fillers, adhesives, pages 48, 114
Synthetic rubbers	Tough, flexible and resistant to chemicals; important components in many coatings, adhesives and fillers. Neoprene, styrene-butadiene rubber, and nitrile rubber are among the best known.	In coatings, adhesives, molding compounds, fillers	Using to make flexible molds, page 78 Using in coatings, adhesives, fillers, pages 48, 114, 116

A plastic for every purpose. The first column of this chart lists the plastics most commonly used in the home. The characteristics listed in the second column are those of the plastic unmodified with fillers or reinforcements, unless otherwise noted. "Tough" refers to resistance to scuffing and wear, "hard" to a lack of resiliency and an unyielding surface, and "strong" to a resistance to tearing and crushing. The third column lists the forms in which each plastic is commonly supplied; it will enable you to see if the material with the right combination of properties comes in the shape or form you need. The last column indexes specialized techniques needed for working with each of the listed plastics.

Storage System for the Raw Materials

A home warehouse in a closet. Store plastic sheets and long rods and tubes on end in a closet subdivided with particle board into narrow compartments. Fill the compartments completely, so that all the plastic is held upright, without bowing or sagging. If you cannot fill a compartment, wedge small cardboard boxes between the sheets or rods and the nearest divider to hold the stock vertical. Short rods and tubes may be laid flat on a shelf, but do not let them overhang the edge, lest the unsupported ends droop or sag. Wrap any unmasked rods in tissue paper to protect them against scratches, and leave the protective masking paper on plastic sheets. Never leave the closet open to sunlight; when masked stock is exposed to sunlight, the adhesive paper can bake on and become difficult to remove.

Liquid resins and hardeners may be stored in the same closet. Solvents, glues and paints should be stored elsewhere to prevent vapors, leaking from the containers, from damaging the surfaces of the sheets and rods.

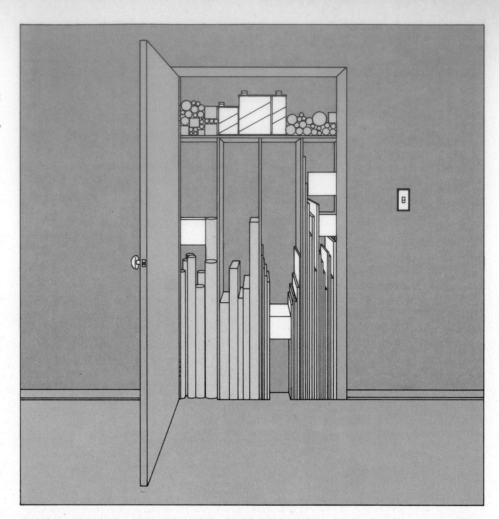

The Unmasking of a Paper-covered Sheet

Stripping off protective paper. Peel back the protective paper from one edge of the plastic sheet and hold it against a wooden rod, such as a broom handle, about 1 to 1½ inches in diameter. Holding the paper securely against the rod, roll the rod across the plastic, wrapping the paper tightly around the wood. Fasten the paper for storage and later use with a strip of masking tape. Do not uncover a greater area of plastic than you need to in order to bend, weld or cement the plastic.

To reuse the protective paper, release the tape securing the paper against the rod, and press the free end of the paper against the unmasked plastic. Gradually unroll the paper across the plastic, smoothing it as you go and pressing it against the plastic surface. At the opposite edge, trim the paper with scissors, and then retape the unused part of the roll.

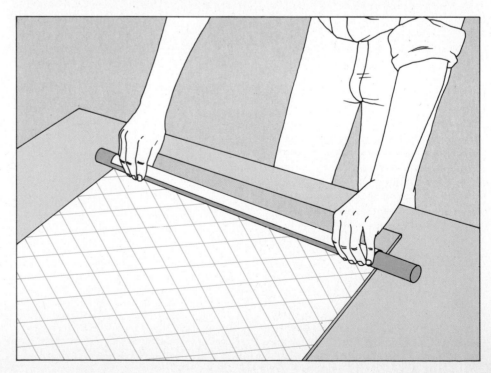

Readying the Stock for Cutting or Drilling

Like wood and sheet metal, plastics can be sawed, drilled and sheared. Also like wood and metal, plastic stock, whether it is flat or tubular, must be marked with lines for layout before the cutting tools are applied.

As they come from the factory or the dealer, plastic sheets are reliably square. So are the ends of rods and rigid tubing. However, if you are using stock that you have cut in the course of previous projects, you should check the squareness of corners, ends and edges. If necessary, true them by clamping the stock in a wooden jig and filing away any uneven plastic. Having squared the edges, you can then calculate cuts that will minimize waste, since straight edges will require no further trimming.

Marking the plastic surface without marring it is the crux of plastics layout: Hard plastics are scratch-prone. Fortunately, most plastic sheets provide a ready-made solution to the problem. At the factory, the sheets are wrapped with protective masking paper; this allows you to indicate cutting lines and locations for drill holes without touching the plastic surface. You can then cut and drill through the marked paper. If you have saved masking-paper wrappings from previous plastics projects (page 14), you can also wrap unmasked rods and tubes for marking.

Only a few tools are required for marking plastics. For straight lines on flat surfaces—sheets or the wider faces of rectangular rods—use a square and either a lead pencil or a china-marking pencil, depending on whether the plastic surface is covered with masking paper. Keep in mind that traces of china marker left on the plastic will bake on if the plastic is heated for bending or welding, so remove the layout lines with a damp cloth after they have served their purpose.

Remember, too, that even a carefully sharpened china marker leaves a line that is fairly broad, although adequate for most work. When your work calls for an especially precise cutting line, or if the plastic surface is so dark that a china marker will not leave a visible line, score the plastic lightly with a scriber. But never scribe marks other than the cutting lines; a scribed fold line or a scribed X

marking a drill hole will remain visible in the finished project and may open the way for a crack. For foam plastics, both stiff and flexible, the standard marking tool is a felt-tipped or ball-point pen; test both to determine which one is best suited to the surface.

When you are drawing or scribing lines on round plastic rods or tubing, a simple V-shaped jig provides both support and control. Two boards, nailed together at a right angle to form a V and propped between two rows of bricks, will serve well. So will an angle iron or a wood beam channeled with a 90° V groove. As shown on page 18, the jig can be used to mark lines around a rod or a tube and also along its length.

Use a compass, with a pencil on one leg, to lay out arcs and circles on paper-masked plastic; use dividers, with two metal points, to scribe an unmasked surface. Before you draw a circle or an arc with a compass or a divider, build up a

foundation of several layers of masking tape at the center of the circle to provide a firm footing for the pivot point, thus protecting the plastic surface from mars.

A more complex shape can be drawn on a separate sheet of paper, then transferred in one of several ways. You can lay carbon paper between the pattern and the masked or unmasked plastic, then trace all the layout lines. Or you can cut out the pattern and fasten it directly to the sheet, rod or tube with water-soluble glue. You can also simply scribe or draw around the cutout pattern.

To avoid scratching unmasked plastic, cover your work surface with a soft pad of newspaper. And to reduce the chance that the completed object will crack or chip, make sure your design includes as few corners sharper than 90° as possible, and no interior angles—they are common starting points for cracks. Curves and broad angles—90° and up—are the safest to use with plastic sheets.

Testing a Cylinder or a Sheet for Squareness

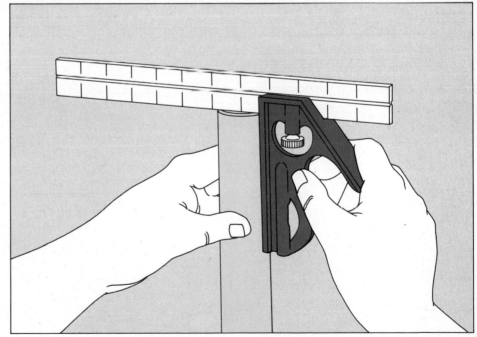

Checking rods or tubes. Fit a combination square or a try square over the end to be tested. Sight between the tool's blade and the stock; if they touch completely while the arm is held firmly against the side of the stock, the end is square. To be sure, move the tool and check the end from an angle perpendicular to the first.

If the end is more than 1/16 inch out of square, use the method shown on page 18 to scribe a true cut line around the plastic, making it as close as possible to the end of the workpiece; then saw off the uneven portion (page 24). If the discrepancy is less than 1/16 inch, simply file the end of the piece until it is square (page 16).

Checking a sheet for square corners. Use a carpenter's steel square to check all four corners of a large sheet of plastic; use a combination square or a try square on smaller rectangles. If you find a square corner, with no space showing between the plastic's edges and the arms of the square, use the adjacent edges to determine corrections needed to square the other corners. If the sheet is more than ⅛ inch thick, make sure the edges are square; swing the arm of the square perpendicular to the sheet, and slide it along the edge, checking several times.

If an edge is less than ⅛ inch out of true, either crosswise or along its length, file it square (*opposite, top*). Otherwise, draw or scribe a new cut line as close as possible to the uneven edge (*opposite, bottom*) and trim the sheet with a saw. If none of the corners are square, correct the one that needs the least work, then proceed as above. To ready an edge for filing, peel the masking paper back and lay a steel square across the plastic. Work an arm of the square toward the edge of the plastic until only the material that must be filed away to produce a square edge protrudes; then scribe a guideline to mark off the waste plastic.

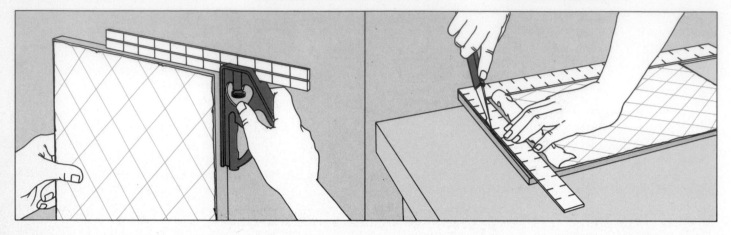

Leveling a Cylinder End

1 Aligning the jig. Select two boards with square ends and with widths greater than the diameter of the plastic rod or tube to be trued. Sandwich the plastic between them. With this assembly resting on a worktable, fit one end in a woodworking vise so that two or three inches extend beyond the front of the vise (*inset*). Use a square to align the board ends, then position the rod or tube so that only the part to be filed away projects beyond the boards. Tighten the vise just enough to hold the assembly; if you are working with tubing, be careful not to crush it.

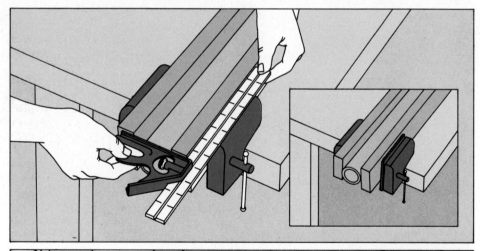

2 Filing the end. Use the boards as a guide while you file away excess plastic with a medium-coarse mill-cut metal file until you have a perfectly square end. On soft plastics such as polyethylene, the file may clog; clean it as needed with the short wire bristles of a file card.

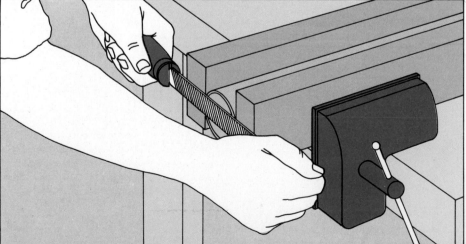

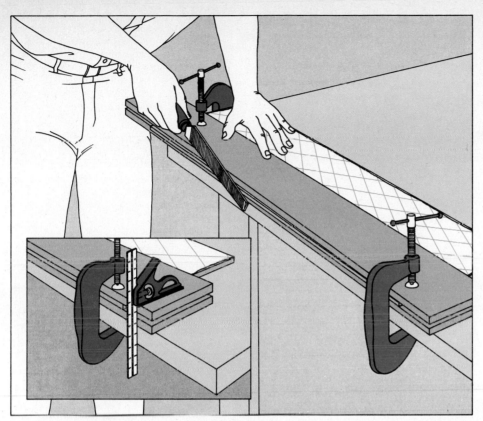

A Wooden Sandwich for Squaring Edges and Corners

Filing away an uneven edge. Sandwich the plastic sheet between a jig made of two boards with straight edges, each a few inches longer than the edge to be trimmed. Align the boards with the scored guideline. To make sure that the boards are evenly placed, fit a combination square over the jig, once at each end; both boards should touch the blade (*inset*). Secure the assembly to the edge of the workbench with two C clamps; if the edge to be filed is more than 2 feet long, use three clamps. With a medium-coarse metal file, remove the excess plastic, scraping away the material with strokes running nearly horizontal along the length of the plastic edge, moving the C clamps as needed.

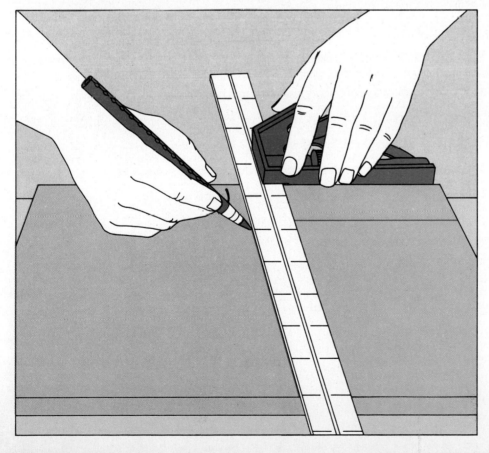

Marking Straight Cut Lines on Sheets and Cylinders

Marking sheets and square rods. Measure and mark reference points for cutting to size with a rule and, depending on the surface, a lead pencil, a china marker, or the point of a scriber. Then align a square or a straightedge with the reference points, and use it to guide the marking tool as you mark the cutting lines.

To indicate the position of holes to be drilled, use a pencil or a marker and the rule to mark two short intersecting lines to form an X with its center at the center of the planned hole. Do not scribe these lines; in most cases they would be longer than the hole's diameter and would be visible as scratches on the completed work.

Girdling tubes and round rods. Lay the tube or rod in a V-shaped jig with a stop at one end; the V groove should be no deeper than one third the diameter of the plastic. Butt one end of the plastic against the stop, and use a ruler or a square to mark the position of the cutting line. Brace a pencil, a china marker, or a scriber against the edge of the jig, pressing the point firmly against the cutting-line mark; then rotate the plastic, extending the cutting-line mark completely around its circumference. Make sure that the end of the plastic stock rests securely against the stop throughout the operation.

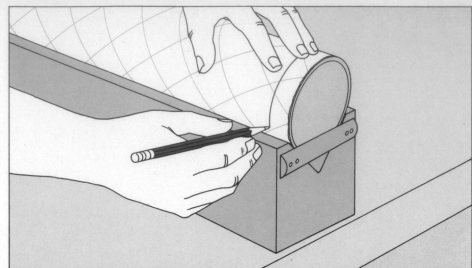

Marking a longitudinal line. Butt the rod or tube against the stop of a V-shaped jig no deeper than one third the diameter of the plastic. Hold the plastic securely with one hand and, using the edge of the jig as a guide, draw a line down the length of the plastic with a pencil, a china-marking pencil or a scriber.

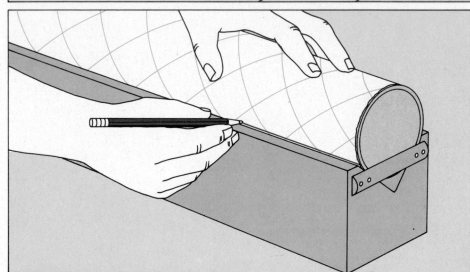

A Padded Center Point for Swinging Curved Lines

A compass or dividers for sheet layout. Set a compass or dividers to the radius of the curve or circle you plan. Without actually touching the plastic, use this radius to locate the center point of the circle; for minimum waste, position the center so that the circle or curve will touch one or more edges of the plastic sheet. Mark the center with a pencil or a china marker, then pad the center with several layers of masking tape. Position the point of the compass or the dividers against the center, forcing the point into, but not through, the masking tape. Then draw or scribe the circle or curve.

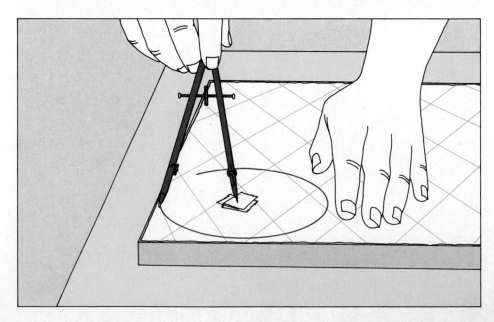

Graph-Paper Patterns for Complex Shapes

A carbon copy for a shelf bracket. Plot the pattern for the bracket on stiff graph paper, using a steel square for the angles and straight lines and drawing the curves either freehand or with a french curve (*inset*). Use the pattern to choose the portion of the plastic sheet that is to be marked and cut, and cover this area with sheets of carbon paper, inked side down. Lay the pattern over the carbon, and secure both to the plastic sheet with masking tape. Trace the pattern onto the plastic, using a ruler for the straight portions of the pattern and drawing the curves freehand. Check to make sure all the lines have been transferred to the bare or masked plastic surface. Then remove the pattern and carbon, and cut the stock.

Make a second bracket in the same way. Glue the two brackets to the underside of a plastic shelf and to a vertical plastic support, which in turn is screwed to the wall. For instructions on gluing plastic to plastic, see pages 48-53.

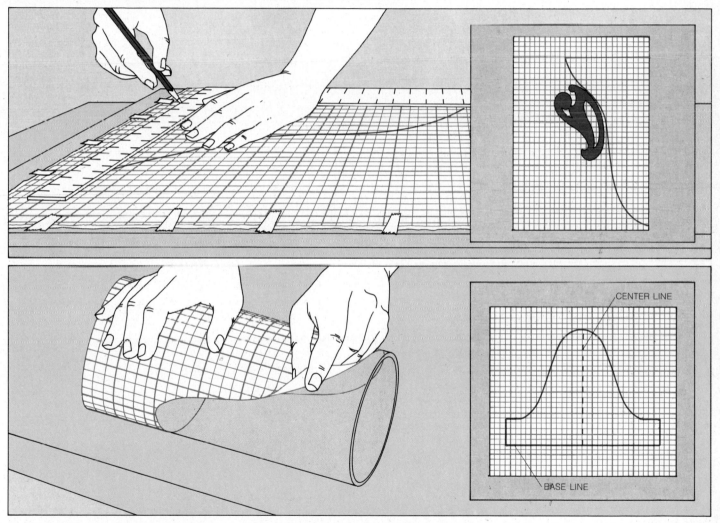

A paper pattern for a plastic scoop. Cut a pattern for the scoop from graph paper and glue it, with water-soluble glue, to a plastic tube of the desired size. In plotting the pattern, locate a base line equal to the circumference of the plastic tube, and a perpendicular center line (*inset*). Then construct the contours of the scoop, using a ruler for the straight sections and drawing the curved sections freehand or with a french curve. The two sides of the scoop should be symmetrical so that their curves can be cut from the plastic simultaneously.

In mounting the pattern on the plastic, line up the base line with the end of the tube and be sure the center line lies along the long axis of the tube. Check to see that the edges of the pattern are well secured. When the glue is dry, cut out the scoop with a coping saw or a band saw, and file the cut edges smooth. Remove the pattern and complete the scoop by gluing a disk of plastic to the end of the tube, then shaping and gluing a handle of plastic rod to the outside face of the disk. For instructions on how to glue plastics, see pages 48-53.

Familiar Cutting Tools, Unfamiliar Techniques

Almost any rigid plastic, from a counter-top laminate to acrylic glass, can be cut and shaped with the same tools used in woodworking. But the tools are not always used in quite the same way. The cutting of plastics calls for particular types of blades, for instance (chart, page 22), and there are some additional safety precautions. When you cut plastics with power tools, make sure the area is well ventilated, and wear eye goggles and gloves. Protection for your hands is especially important when you are working with rough-sawed glass-reinforced plastics or fiberglass. And a respirator is a good defense against particles when you are sawing such material.

In addition, many plastics are themselves in need of protection while they are being cut—because they are so vulnerable to scratching and chipping. Before you run a plastic sheet through any power tool, check the work surface over which the plastic will pass. Even the smallest burr or grain of sand may scratch the plastic. In fact, if the plastic is sold with a protective paper wrapping—commonly the case with acrylic glass—it is best to leave the paper in place while the plastic is being machined.

If the plastic contains no protective wrapping, you can cover it with kraft paper or cardboard, held in place with masking tape or water-soluble glue. To prevent the protective wrapping from clogging the teeth of the saw blade during cutting, lubricate the blade with beeswax or soap.

In lieu of a wrapping, you can cover the work surface with paper, cardboard or a piece of felt. When clamping a scratch-prone plastic, always place a wooden block between the plastic and the jaws of the clamp—whether or not the plastic has a protective wrapping.

During cutting, most sheet plastic will need support to keep it steady. It should rest on a fairly large work surface and be pressed down firmly to reduce vibration. The cut should fall slightly to the waste side of the cutting line, to allow for edge finishing (page 37).

Although power tools cut much faster, almost any handsaw will cut plastic. However, the type of blade used is important. Laminates and acrylic glass can be cut with a standard carpenter's cross-cut saw with fine teeth—10 to 12 per inch. A hacksaw or a keyhole saw fitted with a hacksaw blade will also cut acrylic glass and laminates, as well as PVC pipe and plastic tubing. When small rods are cut by hand, a V block will help to hold them steady (page 18); on round stock, a groove, cut with a three-cornered file on the cutting line, helps to hold the blade during its first strokes.

Of the smaller power tools, the stationary jig saw and the portable saber saw are the most versatile. Not only will they cut all types of plastics but they will handle a range of cuts, such as bevels and curves. Either saw is especially good for cutting intricate patterns or small-radius circles in plastics up to ½ inch thick. However, it is important to fit them with fine-toothed blades when you are cutting fiberglass or glass-reinforced plastics.

When cutting with a jig saw or a saber saw, stop the saw and back it out as soon as it starts to bind. Unfortunately, the heat of a power tool's fast-moving blade tends to melt the plastic and clog the teeth of the saw. It may even fuse together the cut plastic edges behind the blade, trapping the blade in the work.

In general, this problem of clogging and fusing is easier to deal with when you use larger power tools. With table and circular saws, for instance, a wider choice of blades allows you to use a hollow-ground blade with small teeth. Such a blade resists clogging and makes a smooth cut, lessening the tendency of the cut edges to fuse together. With a band saw, the blade has a chance to cool before it reenters the cut.

Another advantage of a band saw is that the plastic chips are dislodged from its blade as it travels around its circuit, resulting in a very smooth cut. However, the wheels over which the blade travels will occasionally need to be cleaned with a wire brush or a file card.

Unfortunately, plastics quickly dull saw blades. For this reason, it may be wise to keep an assortment of blades exclusively for working with plastics. Any blade made for cutting plywood is suitable, if it has the necessary number of teeth per inch, but there are also blades made especially for cutting plastics and so labeled. The blades that stay sharpest are carbide-tipped ones—available for band, circular and table saws—but they also are more expensive than regular blades.

In cutting laminates, especially those not backed with plywood, a major problem is likely to be chipping and cracking, but there are several ways to minimize the risk. One is to press transparent tape over the cutting line. Another is to round interior corners of openings in countertops, such as those for a set-in sink; both during and after cutting, rounded corners are less likely to split than square ones are. A third way to protect the laminate from damage, useful when working with a handsaw, is to place the decorative surface face up and hold the saw at a low angle, moving it in short strokes.

For trimming away excess laminate and shaping a corner at the same time (page 100), a router is useful. It can also cut a V groove in acrylic glass so that the piece, when heated, can be bent to a right angle, forming a sharp interior corner (page 32). As with power saws, if you often use a router for plastics, it pays to invest in a carbide-tipped router bit.

Some plastics can be cut without a saw. For example, acrylic glass up to ¼ inch thick can be scored and snapped in two—but only when the piece removed is at least 1½ inches wide and only when the acrylic is unpatterned, because the bumps at the pattern will cause an uneven score and break. For this technique, shown opposite, you will need a steel square and a ¾-inch dowel, as well as a nail, a linoleum cutter or a special scoring tool designed for plastics. The plastic is snapped while the protective masking paper is still in place, and the paper is cut with a knife or a razor blade.

Small rods and tubing can be cut without a saw, too. First a notch is filed at the cutting line, then the piece is held in both hands, notch facing away from the body, thumbs of both hands together and behind the notch. With a quick motion of the hands toward the body, the plastic is snapped in two.

Thin plastics, such as acetate and very thin laminate, can be cut with paper cutters, tin snips, or large scissors specially designed for the purpose. Such tools are available in plastics-supply stores.

Scoring and Snapping a Sheet of Acrylic

1 Preparing the acrylic for scoring. Mark the break line on the protective paper covering the acrylic sheet, and lay a steel square along the line, clamping both sheet and square to a workbench. Position the break line as far in from the edge of the workbench as the jaws of the clamps will allow. Be sure the positions of the clamps leave the break line free.

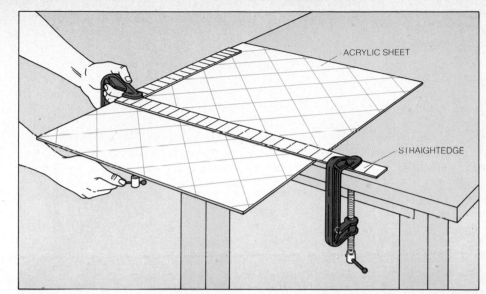

2 Scoring the acrylic. Using the steel square as a guide, score the acrylic repeatedly along the break line with a utility knife or a scoring tool. Apply firm, even pressure along the full length of the line, allowing each stroke to run off the edge of the plastic, onto the workbench. The tool should produce a thin, continuous curl on each pass. Make about 6 passes for sheets that are up to ⅜ inch thick, about 10 passes for sheets up to ¼ inch thick.

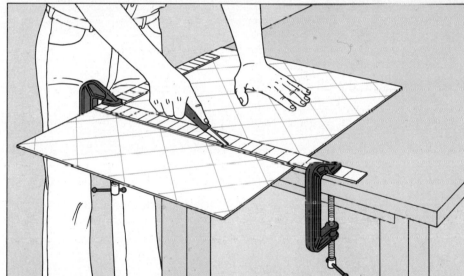

3 Breaking the sheet. Release the sheet and move it to the center of the workbench, positioning the scored line, face up, over a ¾-inch dowel that is at least as long as the line. Brace the wide part of the sheet with one hand and press down on the narrow part, using the dowel as a fulcrum. The sheet should snap cleanly in two.

If the sheet does not break in a single movement, place your hands at the edge closest to you and apply pressure until a break begins. Then move your hands along the line, keeping the heels of your hands about 2 inches ahead of the point where the break ends, gradually elongating the break until the sheet snaps in two. Cut the protective paper with a razor blade or a knife.

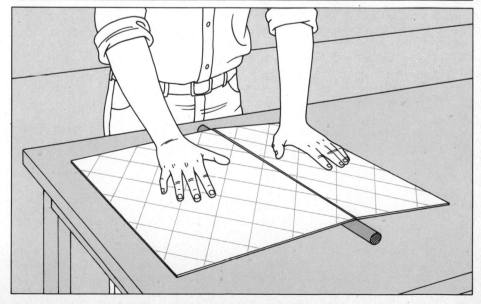

Using Body Leverage to Break a Large, Thin Sheet

Snapping up against a straightedge. Mark the break line on both sides of the sheet, aligning them exactly, and score one side the necessary number of times, as in Step 2, page 21. Place the sheet on the floor, scored side down, wedging two pencils under the outer edge of the section to be snapped. Lay a board on the other section, wide enough to stand on and at least as long as the width of the acrylic sheet. Align the edge of the board with the break line, and secure the board to the acrylic sheet with heavy tape. Then stand on the board and pull up on the free section, snapping the sheet in two.

If the sheet is very wide, have a helper stand on the board while you stand off the sheet and raise it up to break it. Cut through the protective paper over the break with a razor knife.

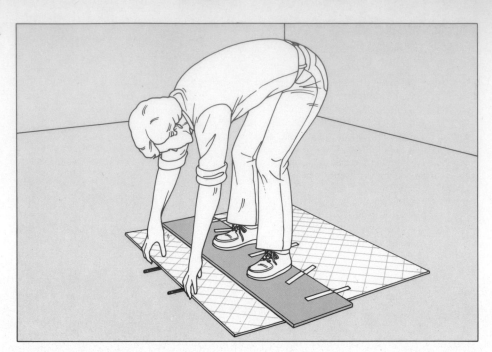

Blade Types for Large Power Tools

Choosing the best blade for the job. The chart at right indicates the best choice of saw blades to use with large power tools for cutting plastics of varying kinds and thicknesses. Listed vertically in two general categories—band saw and circular saw or table saw—are the materials to be cut, the type of blade and the number of blade teeth per inch. In each case the blade listed is the optimum choice, but substitutions are possible. For example, on the band saw, all the listed types and thicknesses of plastics can be satisfactorily cut with a skip-tooth wood-cutting blade. And although a carbide-tipped circular saw blade is ideal for those plastics with which it is paired, a hollow-ground blade for cutting plywood also does the job. The blades listed have odd names but are commonly available. Skip-tooth blades are so called because of a tooth-wide gap between teeth; regular blades for wood or metal have no gap between the teeth. Hollow-ground blades are made with teeth wider than the blade plate, and spring-set blades have teeth that point in alternating directions.

On a variable-speed band saw, adjust the speed to the material. Cut laminates, polystyrene foam and thin plastic such as ¼-inch acrylic at top speed; on home models this is usually 3,000 feet per minute. Tough plastics, such as phenolics and polycarbonate, should be cut at low speed, about 1,500 feet per minute. PVC, cellulosics and thick acrylic should be cut at middle speed, about 2,000 feet per minute.

	Material to be cut	Type of blade	Teeth per inch
Band saw	¹⁄₁₆″-¼″ acrylic	Regular metal-cutting	5-7
	¼″-⅞″ acrylic	Regular metal-cutting or skip-tooth	5-7
	1″ and over acrylic	Skip-tooth	5-7
	Laminates	Regular metal-cutting	3-8
	Up to 3″ cellulosics	Skip-tooth	4-6
	Phenolics, melamine	Regular metal-cutting	6-10
	Epoxies, polyester, acetal, polycarbonate	Regular metal-cutting	10-18
	Polystyrene foam	Regular metal-cutting	10-18
	Polyurethane foam (rigid)	Skip-tooth	4-6
	Polyvinyl chloride (PVC)	Regular metal-cutting or skip-tooth	6-9
Circular saw or table saw	¹⁄₁₆″-¼″ acrylic	Hollow-ground plywood-cutting	8-14
	¼″-¾″ acrylic	Hollow-ground plywood-cutting	5-8
	¾″ and over acrylic	Hollow-ground plywood-cutting or spring-set	3-4
	Epoxies, melamine, phenolics, polyester	Carbide-tipped	8-10
	PVC, polystyrene foam, acetal, cellulosics, polycarbonate	Hollow-ground plywood-cutting	4-6
	Laminate	Carbide-tipped	6-10
	Polyurethane foam (rigid)	Hollow-ground plywood-cutting	4-6

Accurate Band-Saw Cutting with the Aid of Fences

1 Setting up the saw. Install the proper blade (*chart, opposite*) and set the saw for the correct speed; then set the blade guides ⅛ inch higher than the acrylic sheet. Measure the distance from the cutting line, marked on the sheet, to the waste edge of the sheet. Clamp a wood fence this distance from the saw blade. The fence should extend to the near edge of the saw table and 1 to 2 inches beyond the blade. Check the distance from the fence to the edge of the saw table at several points, to be sure the fence is parallel to the table edge and consequently to the plane of the saw blade.

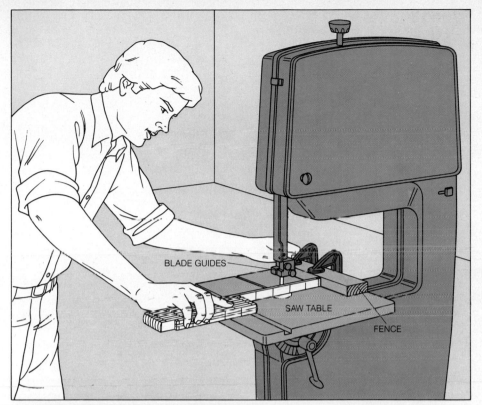

BLADE GUIDES

SAW TABLE

FENCE

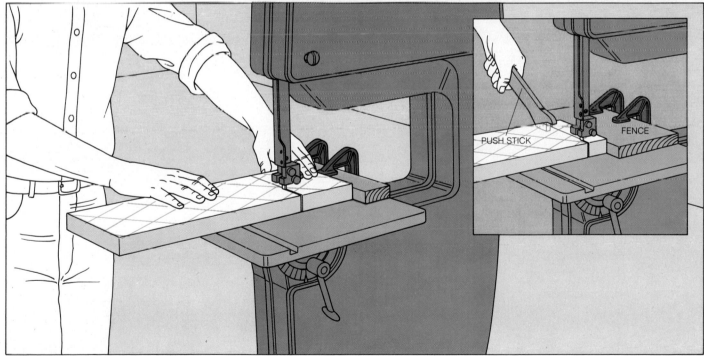

PUSH STICK

FENCE

2 Making the cut. With the saw turned on, butt the acrylic sheet against the fence and push the sheet forward against the blade. The blade should enter the sheet just to the waste side of the cutting line. Feed the plastic steadily past the blade, until the cut is complete. If one part of the stock is less than 3 inches wide, use a push stick to help move the stock along (*inset*).

Forming a Beveled Edge with Power Saws

Using a table saw. Select the correct blade for the plastic (*page 22*) and set it at the desired angle, leaving the blade guard in place; for a smooth cut, the blade should be just higher than the thickness of the plastic. Place the plastic on the saw table, aligning the cutting line with the top of the blade. Butt a wood fence against the plastic to hold it in alignment, and secure the fence with clamps; the fence should extend from the near edge of the table to the midpoint of the blade, but no farther.

Once the cutting position has been established, pull back the plastic and turn on the saw. Then push the plastic past the blade, using the miter gauge as a pusher.

Beveling with a saber saw. Set the saw shoe at the desired angle, and clamp the plastic to a worktable, sandwiching it between the table and a piece of ½-inch plywood. Position the plastic so that the cutting line overhangs the table edge; place the plywood over it so that the distance from the plywood edge to the cutting line equals the distance from the saw blade to the edge of the shoe. Then turn on the saw and begin the cut, pressing the side of the saw shoe firmly against the plywood.

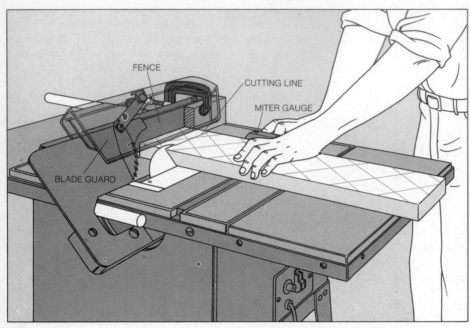

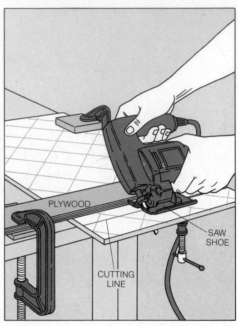

Sawing Tubes and Rods

Using a table saw with round stock. Hold the plastic tube, here PVC pipe, against the face of the miter gauge. To position the stock, hold the plastic against the saw blade, aligning the blade with the cutting line. To cut tubing, turn on the saw, hold the tubing against the miter gauge and the stop, and feed the tubing into the saw blade until the tubing is over the center of the blade. Then begin to revolve the tubing, rolling it toward you, meanwhile holding it securely against the miter gauge. Continue revolving the tubing until you have cut completely around its circumference. File off any rough edges.

To cut a solid circular rod, raise the saw blade until it is $\frac{1}{16}$ to $\frac{1}{8}$ inch higher than the diameter of the rod. Brace the rod against the miter gauge, and push the rod through the saw blade. For a rod greater than 3 inches in diameter, mark the stock, push it past the blade, then turn it over and cut the other side.

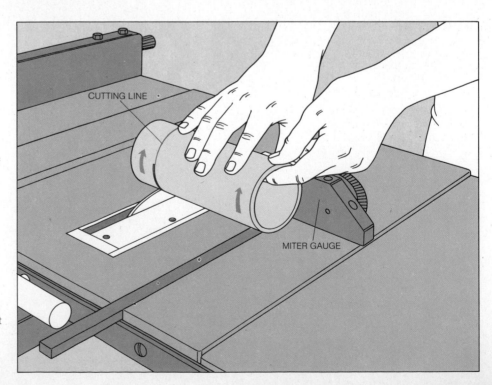

Running round stock through a band saw. Set the blade guides of the saw so that they are ⅛ inch higher than the diameter of the tubing or rod. To prevent the plastic from spinning when it passes through the saw blade, hold it firmly against the miter gauge. Then slowly feed the plastic, here PVC pipe, into the blade until it is cut in two. For rods and tubing of small diameter, you can use a block of wood cut with a V groove to hold the work steady *(inset)*.

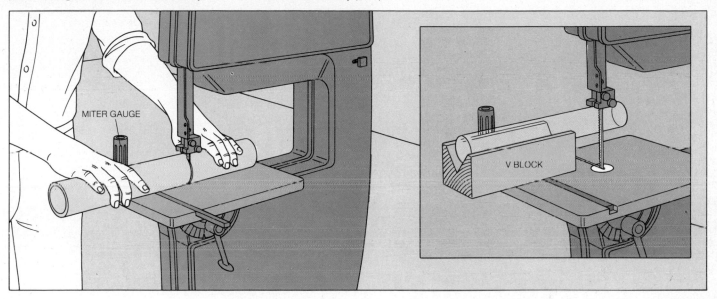

Curve-cutting Techniques

Contouring with a band saw. Prepare the saw by fitting it with the correct blade *(page 22)*, setting it for the proper speed, and adjusting the blade guides to sit ⅛ inch higher than the thickness of the plastic. Turn on the saw and guide the plastic forward until the blade touches the waste side of the cutting line. Then feed the piece steadily against the blade, pushing as fast as the blade will cut easily and stopping only if the blade overheats and sticks.

For an intricate contour, begin with a rough cut, then gradually come closer to the cutting line in several repeated cuts. Use the finishing techniques shown on page 38 to take the piece of plastic down to its final shape.

Shaping curves with a saber saw. Clamp the plastic to a worktable so that part of the cutting line overhangs the table edge (*below, left*). Turn on the saw and cut into the plastic from a waste edge, coming into the cutting line at a shallow angle. Follow the cutting line as far as you can, until the table edge gets in the way. Then turn off the saw, ease it out of the cut and re-

clamp the plastic so that another section of cutting line overhangs the table edge. Reinsert the blade just behind where you left off, switch on the saw and continue to cut. Repeat this procedure until the entire curve is complete.

To cut curves in thin plastic with a saber saw, sandwich the plastic between two slightly larger

pieces of ¼-inch plywood, nailing through the edges of the plywood to fasten the sandwich together (*below, right*). Be careful that you are not nailing into the plastic. Draw the cutting line on the plywood and clamp the sandwich to the worktable, proceeding as above. Use a coarse blade for this job, if you wish, since it has to be able to cut through the plywood.

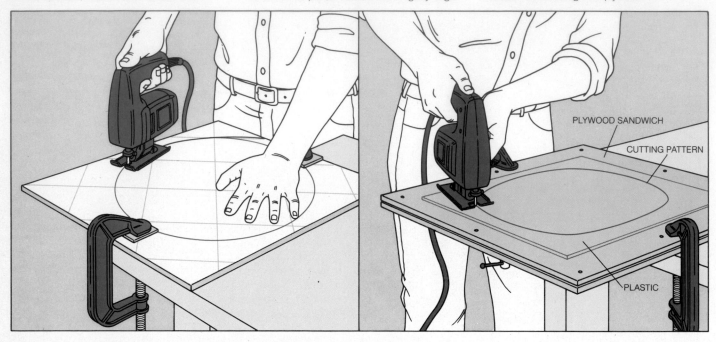

PLYWOOD SANDWICH

CUTTING PATTERN

PLASTIC

Routing a V Groove for a Square Inside Corner

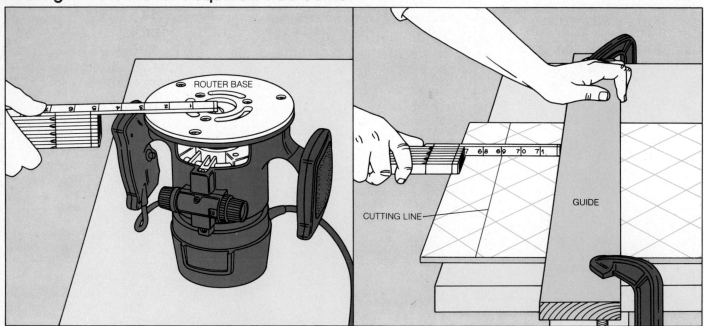

ROUTER BASE

CUTTING LINE

GUIDE

1 Setting up the cut. Draw a cutting line on the protective paper covering the plastic, marking the center of the planned groove, and fit a V-groove bit into the router. Set the tip of the bit flush with the router base and measure the

distance from the bit tip to the edge of the base (*above, left*). Then set a straight-edged guide on the plastic, placing the guide this distance from the cutting line (*above, right*). Clamp the guide in place and recheck your measurements.

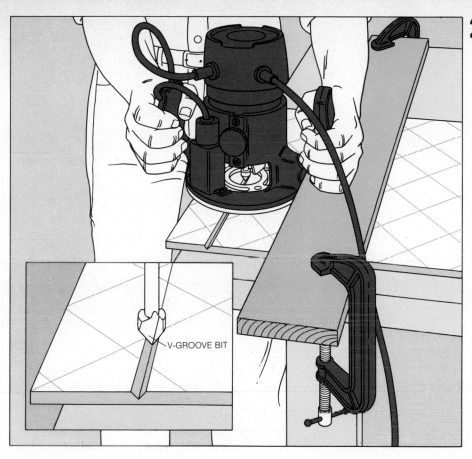

2 **Routing the groove.** Set the depth of the router bit at one half the thickness of the plastic, and place the router at one end of the cutting line so that the bit just clears the edge of the plastic. Turn on the router and draw it along the guide to make a V-shaped groove (*inset*).

V-GROOVE BIT

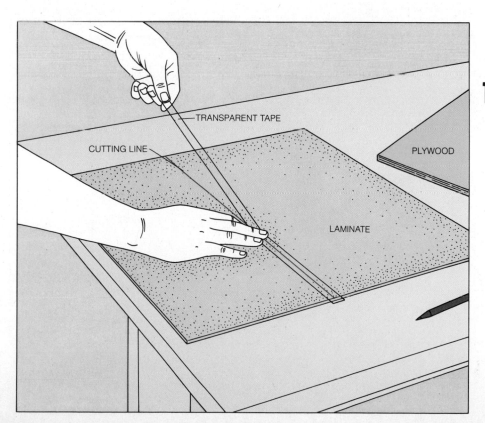

A Professional's Way to Cut Through Laminate

1 **Preparing the cutting surface.** Place the laminate, decorative side up, on a worktable; mark the cutting line, then cover the line with a length of transparent tape. Place the laminate so that the cutting line overhangs the edge of the table by an inch or two. Clamp a piece of plywood over the laminate, positioning the plywood so that its edge can be used as a cutting guide; the plywood will also help to control the vibration of the laminate as a circular saw cuts through it.

TRANSPARENT TAPE

CUTTING LINE

PLYWOOD

LAMINATE

2 **Sawing the laminate.** Fit the circular saw with a plywood-cutting blade that has 6 to 10 teeth per inch, and set the blade depth to 2 inches. Place the saw at one end of the cutting line; then turn on the saw and push it forward, using the edge of the plywood as a guide.

To cut a large piece of laminate in two, set up the work as in Step 1, with the cutting line over-lapping the edge of the table by an inch or two; but with a larger piece you must rest the un-supported end of the laminate on sawhorses or a table about the same height as the worktable.

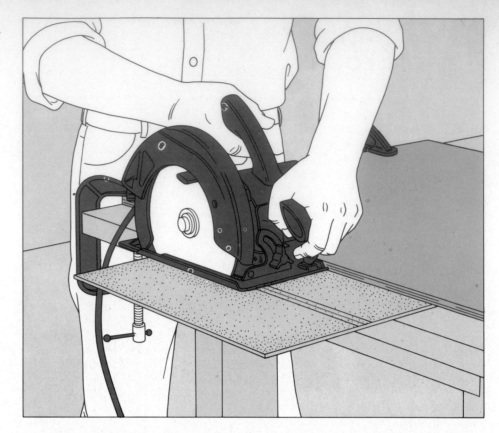

Making an Interior Cut in a Sheet of Laminate

Setting up the work. After marking the cutting line, clamp the laminate, decorative side up, to a piece of scrap plywood, hardboard or particle board and place the work on a pair of 2-by-4s laid across sawhorses. Drill a pilot hole for the saw blade an inch or two inside the cutting line. Insert the saw blade in the hole and cut around the outline.

If the laminate is already glued to a plywood base, as for a sink in a countertop, eliminate the 2-by-4s and the plywood or particle board, and rest the work directly on the sawhorses.

When cutting a glass-reinforced plastic or material that is stiff enough not to need backing but is also brittle—a synthetic marble, for instance—never attempt to cut square corners; they create stress points that may cause the laminate to crack. To round corners with a tight radius, drill a hole at least 1 inch wide at each corner with a hole saw (*page 30*) or a drill, then connect the holes to complete the outlines.

PILOT HOLE

OUTLINE

Cutting Rigid or Flexible Foams

Plastic foams can be rigid or flexible, and they require different cutting methods. Rigid foams such as polystyrene and polyurethane can be cut with most of the same tools and techniques used for acrylic glass *(pages 20-26)*. A jig saw, saber saw or band saw, for example, will make quick work of cutting curved or straight lines in rigid foam. The band saw should be fitted with a wood-cutting or metal-cutting skip-tooth blade having at least 10 teeth per inch *(chart, page 22)*.

A table saw will also cut rigid foam quickly, but it is limited by its blade height to foams less than 3 inches thick. Thin sheets of rigid foam, up to ½ inch thick, can be cut with a razor knife, and very thin sheets can be cut with a paper cutter or with tin snips.

Flexible foams require more ingenuity because they cannot be cut with the power tools associated with woodworking. For thin flexible foam, scissors or a razor blade is usually satisfactory, and an electric carving knife is an excellent tool for cutting straight lines or contours in thicker flexible foam. Another possibility is a commercial foam cutter. Like the carving knife, it consists of two blades moving in opposite directions, but the blades are up to 12 inches long and are attached to a flat base that makes it easier to hold a perpendicular edge.

Rough or uneven edges on rigid foam can be smoothed with a forming tool. Uneven edges on flexible foam are more difficult to deal with, so it is best to make the initial cut as carefully as possible. However, most flexible-foam edges will eventually be concealed—by the upholstery fabric that usually covers them.

Using an Electric Carving Knife to Shape Foam

Making a straight cut. Set the foam on a worktable, positioning it so that the cutting line overhangs the table edge at least an inch. To begin the cut, hold the electric knife at a 45° angle to the top plane of the foam, and draw it into the foam about two inches. Then move the knife to a perpendicular position and continue cutting. Be sure to keep your fingers clear of the blades at all times.

To use a foam cutter *(inset)*, start with the blades upright. Grip the handle, pull the trigger and guide the machine forward—a wheel on the underside rolls the machine along the table.

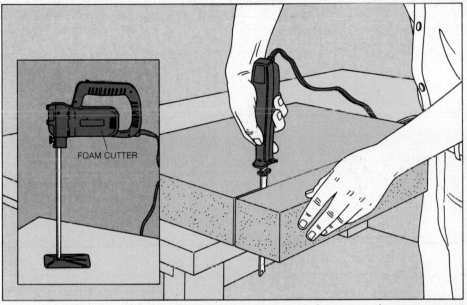

FOAM CUTTER

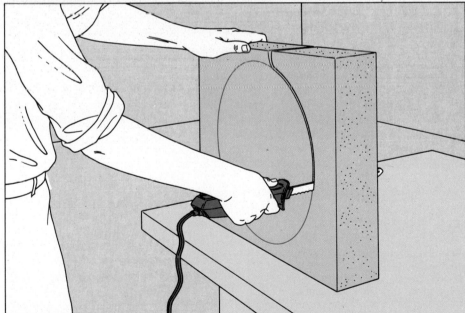

Cutting curved shapes. Stand the foam on edge, cutting outline facing you, and start the cut at the top edge. Follow the outline, keeping the knife blade perpendicular to the foam surface. To make cutting less awkward, you can stop the knife from time to time and change the position of the foam so that you are always cutting with a downward stroke.

For an interior cut, puncture the foam with a sharp knife, and push the electric-knife blade through the slit. Proceed as above.

To cut curved shapes with an electric foam cutter, lay the foam flat on the work surface and guide the machine along the curve, shifting the position of the foam as necessary.

Drilling and Counterboring

Drilling holes in plastics, whether for fasteners or for decoration, is done with the same tools used for drilling holes in wood. But added procedures and precautions are needed, to accommodate plastics' incompressibility, brittleness and low softening temperatures.

The first of these, incompressibility, calls for greater accuracy in drilling holes that must align: Wood fibers will compress to a certain extent when a bolt, screw or plug penetrates two holes that do not precisely match, but some plastics will crack. Where alignment is important, therefore, it is best to use a drill press or to mount a portable drill in a drill stand.

In less critical situations, holes can be drilled freehand with a manual drill or a power drill. When you are using a variable-speed power drill, the speed setting should be in the slow range (500 to 1,000 rpm's) with a high-speed twist bit, and in the high range (up to 5,000 rpm's) with a spade bit. The drill should always be fed into the work slowly and steadily but without excessive pressure, then slowed even more as the tip nears the breakthrough point on the opposite side.

For the actual piercing of the plastic, a high-speed twist bit is preferable, especially when you need to penetrate all the way through stock. But the bit needs modification to keep it from splitting or cracking the plastic as it cuts.

The tip of the bit must be blunted very slightly to keep it from causing small split lines on the other side of the plastic. Except on very thin plastics, the blunted tip permits the full width of the bit to enter the stock before the tip breaks through on the opposite side, stabilizing the drill's action so that it cuts a straight, clean hole. The two sharp leading edges of the bit, called the cutting lips, should also be blunted, so that they plow rather than slice their way along. Some twist bits made for plastics include these blunted features, but it is easy enough to convert an ordinary twist bit with a file.

The configuration of the bit will depend on the size and type of hole needed. A brad-point bit, a spade bit or the drill fitting called a hole saw will work well for drilling plastics. And for countersinking fasteners, you can use the counterbore, countersink and combination drill-countersink bits made for wood.

As further precautions against chipping the plastic, leave the protective paper in place until the drilling is complete. In anchoring the plastic to a work surface or in a vise, use wood scraps to protect the plastic surface from the jaws of the clamp or vise. Working with the plastic at room temperature, rather than when the material is cold, will also discourage chipping.

The low softening temperature of some plastics is both a bane and a blessing in drilling. Bits cutting deep holes should be backed out often to clear the material. Bits should not be stopped in the hole, lest they be seized by plastic cooling around them. On the other hand, the softened plastic lends itself to a technique for producing a clean, smooth hole. Drill a pilot hole slightly smaller than the ultimate hole, and put beeswax or paraffin into it. Then drill again with the proper size bit—the wax lubricates the bit as it spins, helping to expel more chips, and together with the softened plastic, it smooths the walls of the hole.

Finally, when you are working with thin plastics such as acetates and laminates, holes can be punched as well as drilled. Any of the hand punches used for leatherwork will penetrate these plastics.

Drill bits for plastics work. A high-speed twist bit, with its tip slightly blunted, is the best choice for drilling plastics. One designed for wood cutting will drill holes in acrylic glass; a metal-cutting type is best for tough thermoplastics such as nylon, polycarbonate and cellulosics. A brad-point bit, available in diameters of up to 1 inch, is good for drilling holes that go just partway through the plastic.

A spade bit can be used to counterbore holes up to 1 inch wide; its broad paddle provides a convenient surface on which to mark the desired hole depth, and its tip leaves a centering mark for a second bit when you are making a small hole at the bottom of the wide one, for a bolt or screw. It should be very sharp, not modified. A hole saw, available in diameters up to 4½ inches, is useful for large holes in any thickness of plastic. A pilot bit in the center positions the saw on the work, and knockout holes in the sides make it easy to push out the cut disk.

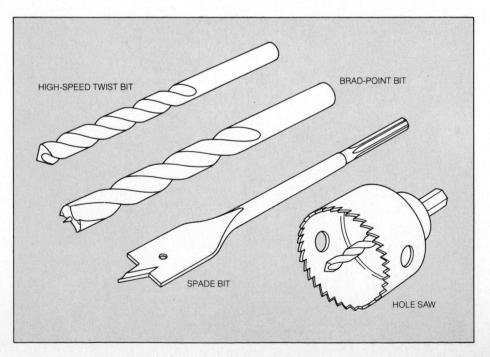

HIGH-SPEED TWIST BIT

BRAD-POINT BIT

SPADE BIT

HOLE SAW

Modified Wood Bits, Special Supports

1 **Modifying the bit.** Round the tip of the bit with a metal file or a bench grinder. Then file or grind the two sharp edges of the cutting lips, making both uniformly blunt. The drawing below shows a twist bit before *(left)* and after *(right)* modification. The cutting angle has been changed so that the drill will emerge without fracturing the plastic. Check the bit by drilling a practice hole in a scrap of acrylic glass. The drill should produce continuous spirals of waste plastic, all the same width. Before drilling, use a punch to make a small dent to hold the tip in place.

2 **Drilling the hole.** Clamp the plastic in a vise between two pieces of scrap wood; the scrap behind the work should be large enough to back up the hole. Turn on the drill and apply even pressure to bore the hole.

On a drill press, back the plastic where the hole is to be drilled with a piece of wood. Center the hole mark under the bit; clamp the wood and plastic to the table, using additional pieces of scrap wood to protect the plastic from the clamp jaws. Drill as above.

For large holes in thin plastic, place scrap wood over as well as under the plastic, and drill through the sandwich. For holes in round stock, use a V block *(page 18)* to hold the rod or tubing, and clamp the V block to the work surface.

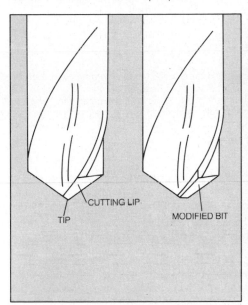

TIP
CUTTING LIP
MODIFIED BIT

SCRAP WOOD

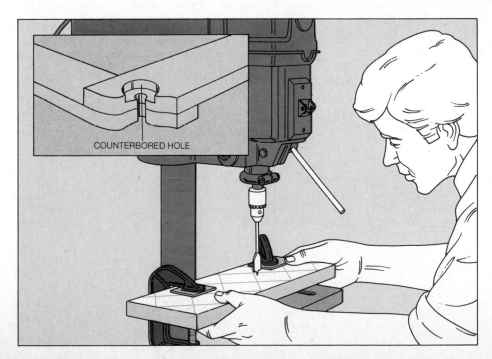

COUNTERBORED HOLE

A Two-Part Hole

Counterboring a hole for a bolthead. Fit a spade bit or a hole saw into the drill press and center the hole mark under the tip of the bit. Clamp the plastic to the drill-press table, protecting the plastic from the clamp jaws with wood scraps. No backing is needed when the hole is being drilled only partway through the plastic. Set the depth stop on the drill press, turn on the drill and lower the bit to make the first cut. Then replace the spade bit with a twist bit $\frac{1}{16}$ inch larger than the diameter of the bolt. Check to make sure the tip of the twist bit lines up with the small hole left by the tip of the spade bit. Put a piece of wood behind the plastic. Reset the depth stop, then drill the second hole.

To drill a matching bolthole in the second piece of plastic, align the pieces, and push a pencil or an awl through the top piece to mark the second one. Set the depth stop and, using the same twist bit, drill the matching hole *(inset)*.

Rigid Sheets and Tubes Made Pliant with Heat

Several plastics have the useful property of becoming pliable when they are heated to temperatures ranging from 250° to 300° F. Once heated, they can be bent into new shapes.

The plastics best suited for heating and bending at home are acrylic glass (commonly sold as Lucite or Plexiglas) and polyvinyl chloride (PVC) pipe. The acrylic, fabricated in sheets of various thicknesses that may be transparent, translucent, tinted, or even mirrored, is often used for making room dividers, break-resistant windows and windscreens. PVC pipe, available in many colors and diameters, is most commonly seen in plumbing disposal systems and drainage fields. After it has been heated and bent, PVC pipe can no longer be used in plumbing, since its walls are usually weakened at the bends. But it is still strong enough to be used for shelf supports or the frames of outdoor furniture.

Sheets of acrylic glass that are ⅛ inch thick become pliable when exposed to 300° heat for about 4 minutes; sheets that are ¼ inch thick need to be exposed to 300° for about 10 minutes. The ¼-inch acrylic is the thickest that is practical to bend at home. PVC pipe softens even faster, in about a minute, at even lower temperatures—around 250°. Neither plastic should be exposed to heat over 300°; both may bubble, scorch and undergo irreversible molecular and structural change if overheated.

Both acrylic glass and PVC pipe cool rapidly and should be shaped quickly while they are at their maximum forming temperature. Moreover, in most cases acrylic should be bent about 5 degrees beyond the desired finished angle, then allowed to retract to that angle, at which point it should be held or clamped for a few minutes of final cooling. Overbending is necessary because the polymers in the plastic (page 97) tend to resist the bend and to return to their former configuration as they cool—a phenomenon known as "memory." When the acrylic is completely cool, this tendency is thwart-

ed, and the polymers are in effect frozen in place. The memory phenomenon has one useful result: It permits novices to correct mistakes and try again simply by rewarming the plastic.

An oven is the most obvious source of heat for bending acrylic glass at home, but the best way to warm PVC pipe to forming temperature is to pour heated sand inside it; this not only softens the plastic but also prevents the pipe walls from collapsing during bending.

Infrared heat lamps may be used to heat large sheets of acrylic glass or isolated sections of a single sheet. In this procedure, the acrylic should be placed on a light-reflective surface, such as sheet metal; shields of the same material placed on top of the plastic can further localize the area to be heated. A special heating iron, sold at paint stores to soften paint for removal, will also do an adequate, though slower, job of heating acrylic to forming temperature, provided the area to be bent is small enough for the iron to heat it all at one time.

One of the most useful heaters for localized areas of acrylic glass is a strip heater; its key feature is a ½-inch-wide tape interlaced with electrical wire to form a flexible heating element. Although commercial models are available, a homemade strip heater (opposite) is less costly and satisfies most home bending needs; the tape is available at plastics-supply stores in 36-inch lengths, complete with the wiring needed to connect it to a house outlet.

Because a strip heater turns a plastic sheet pliable in a very limited area, bending can often be done with only the eye to judge the final configuration. For greater precision, however, the plastic can be bent over a form and locked in place with blocks or clamps.

If the bend line is warmed over a strip heater, the form can be no more than two sides of an angle (opposite, bottom). But if a sheet of plastic is heated in an oven, you can configure it with more complex forms—a coffee can, perhaps, or

the matching halves of a wooden mold cut on a jig saw, or blocks and clamps that hold the plastic in a multiangled shape until it is cool (page 36).

Although acrylic can be bent into an acute angle, the actual apex of the angle will be slightly rounded—unless you adopt a special procedure. A sharp interior corner can be achieved if a V groove is cut along the bend line before the acrylic is heated; the groove should be half as deep as the thickness of the plastic (page 27). When the plastic is bent, the V will form a sharp inside corner; the outside of the corner, however, will be rounded.

Despite the ease with which acrylic glass and PVC pipe can be bent, there are pitfalls. One, to which acrylic is vulnerable, is scratching. To protect the acrylic surface, it is best to leave its masking paper in place except in areas that are to be heated. All paper must be removed, of course, from plastic that will be heated in an oven. But when you use a strip heater, all of the paper can be left in place except for a narrow band.

Another pitfall is underheating or overheating the plastic. When the plastic is bent before it has reached the proper temperature, its internal structure will be strained and small cracks, called crazing, may appear. At the other extreme, overheating may cause the plastic to bulge at the edges of the bend—although these protrusions can be ground down in the finishing process (page 37).

A third pitfall is plastic's tendency to pick up the texture of the material against which it is formed. To prevent this kind of unwanted mimicry, the form can be covered with kraft paper, felt or finely woven cloth.

Of course, no plastic should contact a flame or heated electrical wire, or be left unattended while it is being heated. In addition, when an oven is used for heating, its door should always be left slightly ajar so that potentially toxic fumes released from the plastic will not build up inside, and the kitchen should be adequately ventilated.

A Simple Heating Tool for Bending a Narrow Strip

Anatomy of a strip heater. A homemade strip heater, shown assembled in the inset at right, is a layering of wood, foil and insulation paper into which a band of electric heating tape is placed. The base is a length of ½-inch-thick plywood 6 inches wide and about 6 inches longer than the heating tape. Nailed on top of the base are two ¼-inch-thick plywood strips set ¾ inch apart to form a shallow channel as long as the heating element. Covering the strips and the space between them are two layers of heavy-duty aluminum foil. At one end, the foil is pierced by a screw for a ground wire, which should be long enough to reach a ground connection such as the cover-plate screw on an electrical outlet.

Placed over the foil is a double layer of high-temperature insulation paper, normally used as an oven liner and available in hardware stores. The paper is scored or dampened to follow the contours of the channel and stapled in place, along with the foil, against the plywood. The heating tape is centered within the channel, pulled taut and tied with its strings to pairs of nails tacked near the ends of the plywood base.

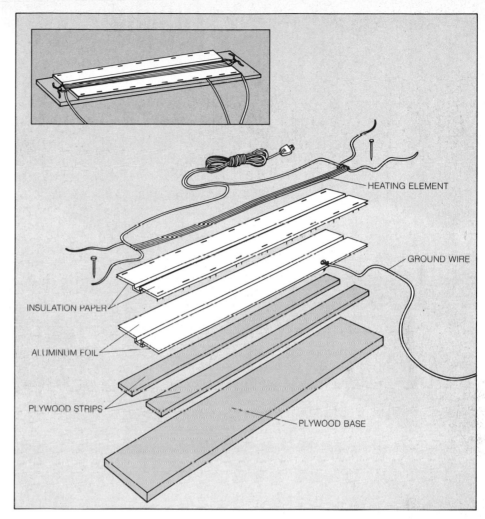

HEATING ELEMENT

GROUND WIRE

INSULATION PAPER

ALUMINUM FOIL

PLYWOOD STRIPS

PLYWOOD BASE

Forming an Angle with a Homemade Strip Heater

1 **Fashioning a cardboard guide.** Score and fold a piece of heavy cardboard into the desired angle, anchoring it with masking tape against a second piece of cardboard. Check the angle with a protractor, and adjust the form if necessary.

2 **Preparing the acrylic.** Measure and mark the location of the bend line on the two edges of the plastic (*below, left*), using a grease pencil or a china-marking pencil. Measure out 1½ inches on each side of these marks, then draw two pencil lines across the face of the acrylic, still covered with its protective masking paper. Score the masking paper lightly along the lines, using a dulled glass cutter so that you do not mar the surface of the acrylic. (To dull the blade of the cutter, run it back and forth several times over a piece of metal.) Exerting just enough pressure to score only the paper takes practice; try it on acrylic scraps first. Finally, peel off the 3-inch strip of masking paper between the lines, along with its adhesive, and remove any traces of the adhesive with the recommended petroleum-base solvent (*page 48*).

For a precise 90° angle, groove the bend line with a router's V bit. You may polish the faces of the V by sanding them, then wiping with a cloth soaked in methylene dichloride, but observe the procedures and cautions on pages 41 and 48.

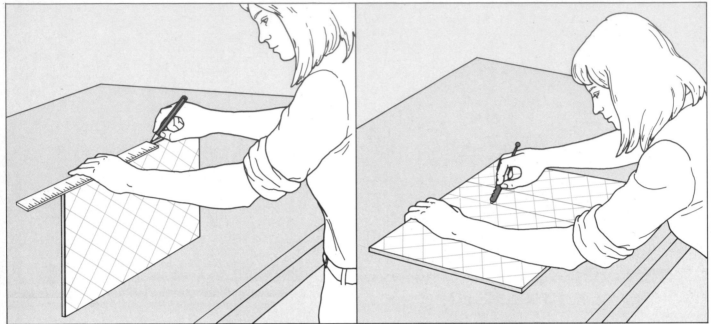

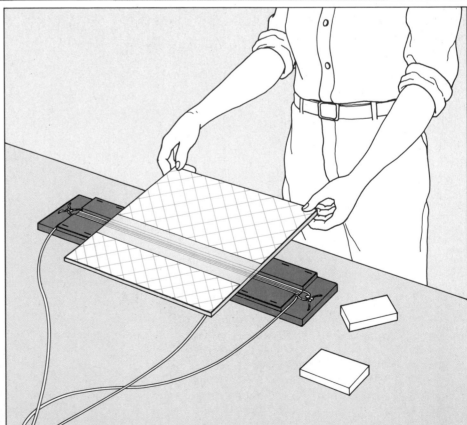

3 **Heating the acrylic.** Position the acrylic on the strip heater so that the two marks for the bend line are centered over the heating tape. Weight the acrylic with metal blocks to keep it from shifting position and, with a soft cloth, rub off the bend-line marks. Plug in the strip heater and warm the plastic until its edges droop slightly.

4 Bending the acrylic. Remove the weights from the acrylic on the far side of the heater. Hold down the near side of the acrylic with one hand while you grasp and gently pull the far edge toward you, bending it to the desired angle and then about another 5° beyond (*below,*

left, inset). Remove the acrylic from the heater, set it on the workbench and allow the bend to retract. Then, while the acrylic is still warm, hold it over the form—the one shown here is in the shape of a tent—until it has cooled, about a minute (*below, right*).

To bend an acrylic sheet that has already been grooved, set it in position over the heater so that its grooved side is facing up. Bend the warmed acrylic sheet only until the two faces of the V meet; overbending is not necessary in this case (*below, right, inset*).

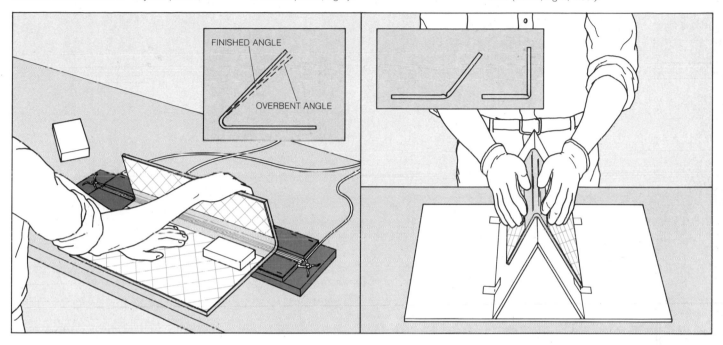

FINISHED ANGLE

OVERBENT ANGLE

The Softening Influence of a Heated Oven

1 Heating in an oven. To soften a sheet of acrylic to a rubbery consistency, remove its masking paper and place it on a clean, flat cookie sheet in a preheated 300° oven. Leave the oven door slightly open so that any fumes released by the plastic can escape, and ventilate the kitchen. Acrylic that is ⅛ inch thick will soften in about 4 minutes, ¼-inch-thick acrylic in 10 minutes.

To guard against overheating, check the oven controls first with an accurate oven thermometer.

2 Forming the sheet. Curve or twist the softened sheet into the desired shape, working by eye or using a prepared form. For a cylindrical shape or a half cylinder, hold the sheet over a coffee can or an oatmeal box (*right*), padding the can or box with felt to prevent the acrylic from mimicking irregularities.

For compound angles, such as those needed to form a lipped shelf, clamp the plastic between blocks and wedges (*inset, top*), keeping in mind that the corners of the blocks must be slightly beveled to accommodate the plastic's rounded corners. For an undulating shape, clamp the plastic between the two matching halves of a wooden block that has been cut into a male-female mold (*inset, bottom*).

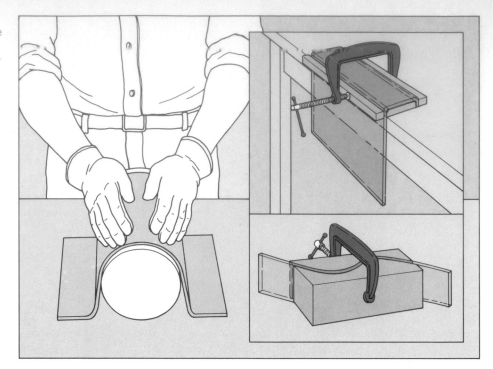

Bending PVC Pipe into Shape with Hot Sand

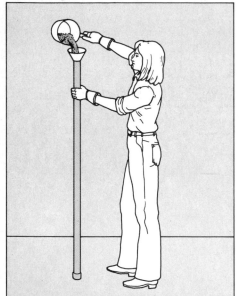

1 Filling the pipe with sand. To warm the sand, place it in an old pot over low heat on the kitchen range. Stir it occasionally to distribute the heat, and check its temperature every few minutes by putting the end of a scrap of PVC pipe into the sand. When the sand is warm enough to make the PVC pliable, temporarily cap one end of the pipe to be formed and pour sand into the other end, through a funnel. Cap the pipe when it has been filled. Depending on the weight of the sand within the pipe, you can use regular PVC caps or several layers of aluminum foil secured with rubber bands.

2 Bending the pipe. Test the sand-filled pipe for flexibility, and as soon as it is easy to bend, shape it by hand or around a form. To hasten the cooling process, have a helper remove the caps; then you can empty the sand out or flush it out with a hose. Spraying cool water on the pipe will also help shorten the setting time.

Giving Luster and Polish to Edges and Surfaces

Most decorative plastics come fully polished, protected by paper masking. But unmasked edges must be smoothed, as must any cut or scratched surface.

This smoothing process consists of scraping, filing and sanding to remove rough edges and tool marks, followed by buffing and polishing to bring a high gloss to the surface. Not all surfaces need to receive the full treatment. The edges of a tabletop or a shelf should be buffed and polished until they are crystal clear, but edges that are to be glued or welded can simply be smoothed and left opaque.

The first step, scraping, applies especially to hard plastics, such as acrylics and polycarbonates. A refinishing scraper is the most efficient tool, but you can improvise by grinding down an edge of a triangular file. Some softer plastics, such as polyvinyl chloride (PVC), can be smoothed with a block plane.

Along inside edges and interior cutouts where a scraper cannot reach, filing is the principal finishing technique. Woodworking tools, such as smooth-cut rasps and bastard-cut mill files, are best for finishing edges; half-round, triangular and curved riffler files are useful for tight curves and fine details. A piece of chalk rubbed over the blade will keep the file from sticking.

Sanding, the next step, is best done wet because the friction-caused heat of dry sanding can soften most thermoplastics. Abrasive papers for plastics are available in hardware stores. Steel wool can be substituted for abrasive paper.

Mechanical sanding with an orbital or a belt sander is faster and easier than hand sanding. To avoid heat build-up, keep the sander or the plastic in constant motion, avoid extreme pressure and frequently wet the surface being sanded.

Like sanding, buffing can be done either by hand or mechanically. To buff by hand, use either a shoeshine buffer or a piece of soft flannel wrapped around a wood block. When buffing irregular forms or curved surfaces, use a strip of flannel in the same way a shoeshine cloth is used. Buffing compounds of the type commonly used on metal are also suitable for plastic. They come in several grades, colored according to coarseness. The finest of these compounds, usually white, are the coarsest you will usually need for plastics.

Mechanical buffing can be done on any kind of buffer. You can adapt a double-shafted bench grinder for buffing by substituting loose cotton buffing heads for the grinding wheels. For large sheets, use either a portable electric shoe buffer or an electric drill fitted with a buffing disk. Coat, or charge, the buffing head by holding a stick of buffing compound against the rotating wheel or disk until it is well covered with compound.

Any good-quality floor wax or car wax can be used to polish plastics, but plastics manufacturers often recommend specific compounds. These polishing pastes are applied in exactly the same manner as the buffing compounds.

In thermoplastics, such as acrylic and polycarbonate, drilled holes can be polished with methylene dichloride solvent, available in solvent-cementing kits (page 48). The solvent, poured into the hole, works on the plastic by dissolving the thin frosty layer. This solvent can also be applied with a cloth to smooth and polish the sides of a V groove.

Shallow scratches on the face of plastics can be polished out with white toothpaste. If the scratch is deep, sanding may be necessary, but because optical distortion may result, clear acrylic windows or skylights should not be sanded.

Abrasives Useful for Sanding and Buffing

Abrasive	Recommended coarseness			Comments
	Rough	Medium	Fine	
Sandpapers				
Garnet	100 grit	180 grit	280 grit	Durable
Silicon carbide	80–100 grit	180–220 grit	330–600 grit	Fast-cutting, durable
Tungsten carbide	36 grit	80 grit	150–200 grit	Very durable, can be cleaned with a wire brush or methylene dichloride solvent
Steel wool	No. 0	No. 00-000	No. 000-0000	Can be used instead of sandpaper
Buffing compounds				
Rouge	not used	not used	white	Available in stick or cake form
Tripoli	not used	not used	white	Cuts slightly faster than rouge

A full range of abrasives. Abrasive materials commonly used on plastics are listed in the column on the left. Listed to the right of each abrasive is the suitable range of grit or grain for rough sanding, semifinishing and fine finishing. All of the sandpapers listed can be used wet, and for plastics they should be. Always start with the finest grade, or highest grit, abrasive paper so that the surface of the plastic will not be roughened more than necessary.

Step-by-Step Smoothing of Tool-Roughened Edges

1 **Scraping to remove saw marks.** Clamp the plastic in a vise, protecting the masked faces from damage by inserting scrap wood between the plastic and the vise jaws. Grasp the scraper with both hands and position it across the stock, tilting it back toward you at an angle of 60°. Starting at the far end, drag the edge of the scraper over the length of the stock repeatedly, from back to front. Never push the blade forward. Use moderate pressure and long strokes to avoid creating depressions in the piece.

2 **Wet sanding.** Fit a slab of plate glass or wood into one end of a shallow metal pan, and add water to a level just below the surface of the slab. Lay a full sheet of 100-grit wet-or-dry sandpaper on the slab, grit side up, and wet the paper. Dip the plastic in the water and move the edge over the paper in an oval pattern. Keep the edge flat on the paper and use even pressure. At frequent intervals, wet the plastic and rinse off the abrasive paper.

When the surface is even and smooth, repeat the process with progressively finer grades of paper (*chart, page 37*) until the sanding marks are almost undetectable. Remove the fine dust by rinsing the plastic under running water.

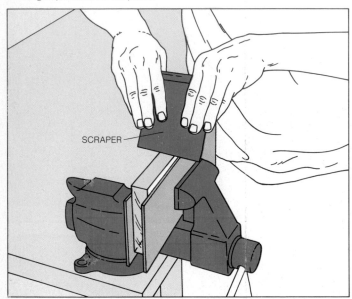

3 **Buffing and polishing.** To buff an edge, lift one end of the plastic against the bottom quarter of a buffing wheel coated with white compound. Keep the stock in constant motion, moving it from side to side and up and down. Buff the edge up to its center point, then reverse the work to buff the opposite end.

Wash the stock with soap and water to remove the abrasive buffing compound. Then polish the edge by coating the second buffing wheel with polishing paste or paste wax, and move the edge over the wheel, as above. Continue polishing until the edge is shiny, and then wipe the edge with a clean, soft cloth.

To buff and polish an edge by hand, clamp the plastic in a vise between protective wood scraps (*insert*), and use a sheepskin shoeshine buffer coated with the same compound and wax as above; use separate buffing heads for the two operations. Hold the buffer flat against the edge and move it back and forth, parallel to the edge. Shift the position of the plastic often, to smooth the edge evenly.

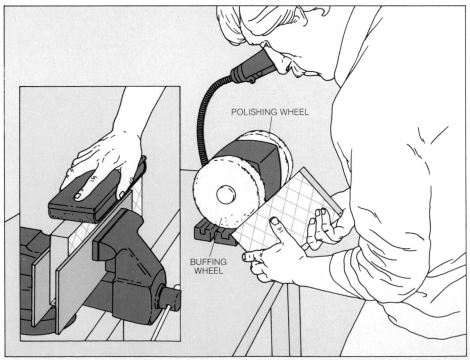

Smoothing and Buffing an Inside Curved Surface

1 **Filing with a half-round file.** Clamp the plastic in a vise between protective wood scraps and, gripping a single-cut, half-round file with two hands, file across the plastic at right angles to the edge. Use light, even strokes and a forward motion, twisting the file a half turn on each stroke. Rotate the plastic in the vise as needed to keep the surface that is being filed just above the jaws. Use a file card to clean the teeth of the file after every few strokes.

2 **Sanding and buffing.** Wrap a piece of 100-grit wet-or-dry sandpaper around a dowel smaller in diameter than the curve to be sanded. Wet the sandpaper and the plastic with water, then sand with a combination back-and-forth and rolling motion, keeping the dowel level with the edge and working slowly around the circle. When the edge is fairly smooth, change to 240-grit, then to 400-grit paper. Rinse the plastic well with water each time you change the paper.

Replace the abrasive paper with felt. Coat the felt liberally with white compound and buff the edge, using the same technique as in sanding. Buff until smooth, then wash off the buffing compound with soap and water.

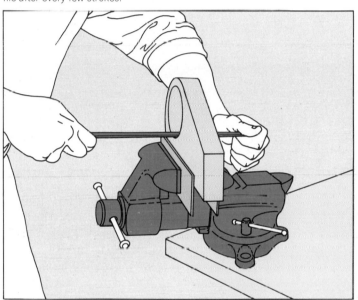

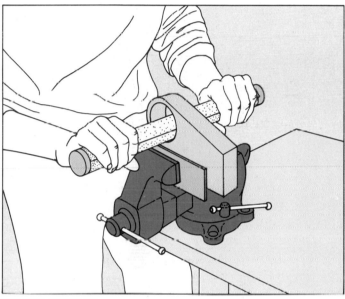

3 **Polishing to a shine.** Saturate a strip of lint-free cloth with polishing paste or paste wax. Holding the ends of the cloth firmly, draw it briskly back and forth over the surface. Polish until the surface shines, then repeat the procedure with a clean, dry piece of cloth.

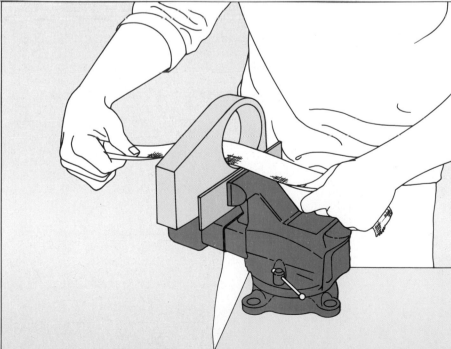

Removing a Scratch from an Acrylic Sheet

1 Sanding the sheet. Lay the plastic on a piece of felt on a flat work surface, and wet the plastic with water. Using a sanding block and a fine-grade wet-or-dry sandpaper (approximately 280-grit for shallow scratches), sand very lightly in a star pattern. Keep the sheet and the sandpaper wet, and work over a wide area to avoid creating a noticeable depression.

As the scratch disappears, use progressively finer grades of wet-or-dry paper in the same star pattern until the surface is uniformly smooth. Wash off the dust with soap and water, and dry the plastic with a damp chamois.

2 Buffing and polishing. Use a portable electric drill fitted with a wool buffing disk, and coat the disk with buffing compound. Keeping the disk in constant motion, buff the sheet in small, overlapping loops until the sanding marks have disappeared and the stock is slightly shiny. Wash away the buffing compound, using your bare hand and soap and water.

Polish the sheet in the same manner as above, using a clean wool buffing disk, and either paste wax or polishing paste. When the sheet has the desired luster, finish the job by hand-polishing it with a clean, dry cloth.

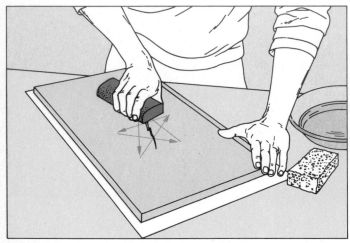

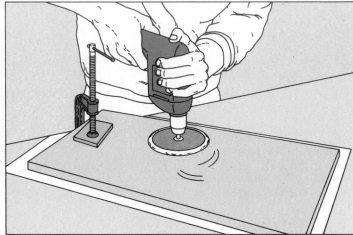

Two Ways to Polish a Small-Bore Hole

Using a drill stand. Cut a piece of wooden dowel, slightly longer than the depth of the hole to be polished and 1/8 inch smaller in diameter, and cut a 1/2- to 1-inch slit at one end. Slip a small piece of 280-grit wet-or-dry sandpaper into the slit (inset) and tighten the dowel in the drill chuck. Clamp the plastic, supported by a piece of scrap wood, against the drill-stand table, centering the hole under the dowel. Squeeze a few drops of lubricating oil into the hole, then raise and lower the rotating dowel into the hole until the inner surface is smooth.

Clean the hole, then repeat the process, substituting a strip of felt, coated with white compound, for the sandpaper. Clean the hole again and repeat the process, using a new piece of felt coated with paste wax or polishing paste.

The same results can be accomplished by hand if you simply twist the dowel between your thumb and your forefinger, at the same time moving it slowly in and out of the hole.

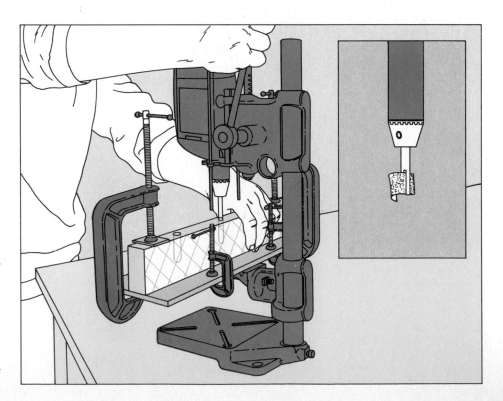

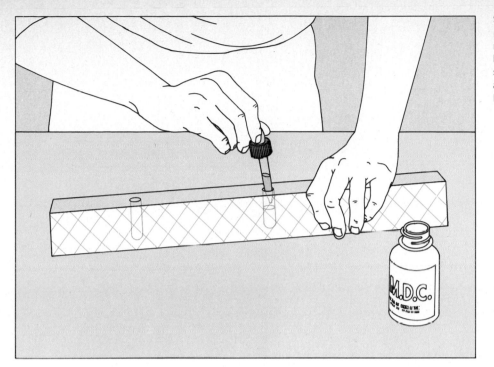

Solvent polishing. After knocking any loose particles or dust from the hole, hold the plastic upright and use a medicine dropper to fill the hole to the brim with methylene dichloride. Let the solvent work for 30 seconds, then pour it into a disposable container. Flush the hole immediately with soapy water; rinse with clear water.

The Light Touch That Cleans Plastics

For all their toughness, plastics require a surprisingly gentle hand when it comes to maintenance. Thermoplastics especially acrylic, polycarbonate and polystyrene—can be easily scratched, and because these plastics are usually clear, scratches caused by rough cleaning will be highly visible.

To remove superficial dirt from a thermoplastic, rinse the surface first with running water to rid the surface of any loose particles, which can scratch. Then wash with soap or a mild, non-abrasive detergent and water, using your bare hands or a soft cloth. Do not wipe with a completely dry cloth, since doing so can generate an electrostatic charge that attracts more dust and dirt. Antistatic solutions called demagnetizers, available from plastics suppliers, negate this charge and also serve as a protective polish to fill in shallow scratches and prevent future ones.

Avoid abrasive scouring compounds, boiling water and harsh solvents such as acetone, benzene, toluene, tetrachloride, and lacquer thinners—all of which will attack thermoplastics. So will many window cleaners. Isopropyl alcohol will remove the adhesive residue sometimes left by the paper masking on thermoplastic sheets. And if a solvent is needed to cut grease on these surfaces, use a good-quality kerosene or naphtha; remove the solvent immediately with soap and water, and dry the surface with a damp chamois. To remove yellowing from polycarbonates, buy butyl cellosolve from a plastics supplier and repolish the surface with a polishing paste or paste wax, buffing it back to its original luster with a soft cloth.

Thermosetting plastics, such as the laminates used on kitchen or bathroom countertops, are not as sensitive as the thermoplastics are to solvents or abrasives, but they can be scratched with sharp objects and will discolor if subjected to high heat. Do not try to remove spots with metal scrapers or scouring pads; wipe with a wet sponge and soap or a mild scouring compound. Polish with a clean cloth.

Fiberglass-reinforced plastics have a tendency to craze, or develop fine surface cracks, when attacked by harsh solvents. Washing with detergent and water will remove most dirt; use isopropyl alcohol or kerosene when a solvent is needed to clean tough stains.

Welding a strong joint. An invisible jet of hot gas from a plastic-welding tool melts a PVC filler rod between the edges of two PVC sheets. The edges of the sheets have been beveled to form a V joint. The joint edges and the rod are simultaneously heated with an oscillating motion of the welding-tool tip. As they soften, the rod is pushed gently into the joint, fusing the plastic surfaces. Several rods will be melted along the joint, overlapping each other until the V is filled.

The methods for joining one piece of plastic to another are both reassuringly familiar to and startlingly different from methods for joining metal and wood—a reflection of the chameleon nature of these synthetic substances. Many plastics can be joined with mechanical fasteners as prosaic as stove bolts and rivets; others with adhesives as mundane as white glue. But sometimes neither of these two standard fastening techniques is as effective or as convenient as one of several methods peculiar to plastics.

Two of these methods take advantage of plastics' most vexing weaknesses: their sensitivity to heat and to strong cleaning compounds. Many a homeowner has discovered that polyethylene food wrapping instantly melts if brushed against a hot saucepan. But that same sensitivity to heat makes it possible to weld plastics at temperatures far below those of conventional metal welding, using a hot plate or a household iron for a welding tool.

Similarly, the solvents in window-cleaning compounds that soften the surface of shatterproof plastic glass, causing it to haze and crack, also yield a method for creating an invisible joint. Used in small quantities and in confined areas, certain solvents soften two plastic edges so that, when pressed together, their molecules mingle, then harden into a molecular bond.

Not all of these fastening methods, familiar or unique, will work on all types of plastics. The world of synthetics is too varied. There are some plastics, especially thin films, that are too soft to accept mechanical fasteners; others, such as polyethylene, have slick, nonporous surfaces that reject adhesives, causing them to peel right off. Still other plastics, notably the polystyrenes, char when heated, making hot-gas welding impractical. Choosing the right technique becomes even more crucial in joining two dissimilar plastics or in joining a plastic and, say, wood.

Narrowing the choice still further are practical concerns. You will want to consider the conditions under which the assembled parts must perform. Must the joint be watertight or airtight? Solvent-cementing and welding are fine, but screws and rivets will not do. Must the joint be flexible enough to expand and contract with extreme temperature changes? Use a silicone adhesive and avoid white glue. Do looks count? The most attractive joint is produced by solvent cement, the least attractive by hot-gas welding. But a welded joint is rugged. Will the parts need to be disassembled? Nuts and bolts are the logical choice. These and other considerations are discussed in the chapter that follows, making it possible for you to join plastic parts by the method that best suits the physical or esthetic demands of the project.

Mechanical Devices: Joinery's Nuts and Bolts

The mechanical fasteners often used on plastics are the familiar screws, rivets, nuts and bolts that join wood and metal. Many of their applications are familiar too, but the unique properties of plastics allow for some variants on joinery techniques: Screws can be set into threads heat-formed into thermoplastics; nuts can be embedded in plastic filler for a strong, invisible joint. Many plastics are virtually friction-free, a property that makes them useful for simple smooth-swinging hinges in which ordinary long brads serve as hinge pins.

Rivets are the fasteners of choice for connecting thin sheets of plastic to each other or to nonplastic materials. Although they can be drilled out of their holes if necessary, rivets are best suited for permanent installations. They are also good for applications where vibrations might cause a threaded fastener to work loose. Screws or nuts and bolts provide greater strength than rivets and can be used with materials of any thickness. Most of them allow assemblies to be dismantled repeatedly. An exception is the thread-cutting screw, whose repeated cutting action in screwing and unscrewing can damage threads cut in plastic.

Adapting Traditional Fasteners to New Uses

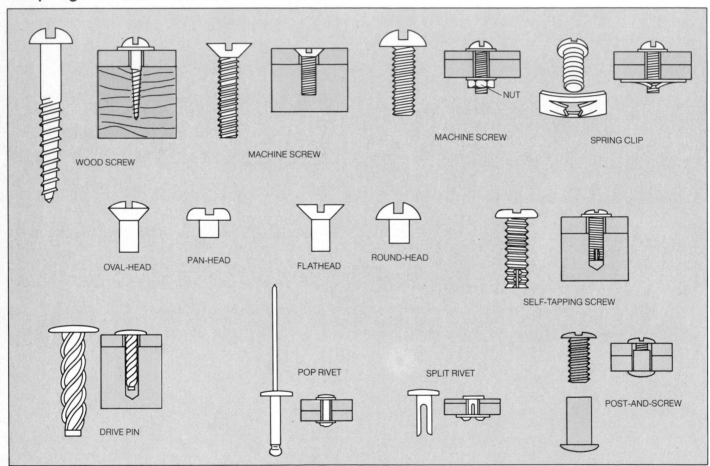

Fasteners for every application. Wood screws are best suited for connecting plastic to wood, and they can be heated to form matching threaded holes in thermoplastics (*page 46*). Machine screws can be used to attach plastic to any material, but they need a matched fitting, either a prethreaded hole cut with a tap (*page 47*), or a standard nut or a spring clip (which is pushed rather than threaded onto the screw). Both wood screws and machine screws are available with flat or oval heads that lie flush with the surface in countersunk holes, or with round or pan heads that protrude above the surface and are generally used with washers to help distribute the pressure more evenly.

Self-tapping screws and drive pins, both originally designed for metal, cut their own threads. Self-tapping screws carve their own path as the screwdriver turns; drive pins are forced into their holes by the pressure of hammer blows. Self-tapping screws can be used in both thermoplastics and thermosetting plastics, but drive pins should be used only in thermoplastics, as they may cause cracks in the relatively non-yielding thermosetting plastics.

Rivets can be used to join plastics to plastics or to other materials. Pop rivets are installed with a special tool that expands the tip of the rivet shaft into a head; split rivets are set by having their split ends bent outward with a screwdriver. Both are inserted in predrilled holes and should be used with washers, to distribute pressure.

Combination post-and-screw fasteners come in a variety of lengths and can be used to join two pieces of plastic where little structural strength is required. The hole in each piece should be drilled to the diameter of the post.

With any such fasteners, the physical characteristics of plastics give rise to other problems. Most plastics are not as strong as wood or metal, so you must avoid excessive strain at any point. Tension on threads cut into plastic can strip the threads; tighten fasteners only until increased resistance is felt. Space fasteners closely to distribute strain evenly.

To further distribute the pressure of each fastener, place a broad washer between the head of the fastener and the plastic. In many industrial designs a simi-lar result is obtained by means of a thicker section, called a boss, molded into the plastic part around each fastening point. If you are molding a plastic part, you can do the same.

Because most plastics are more responsive to temperature changes than other materials are, you must leave room for expansion and contraction when you attach a piece of plastic to another material at more than one point. If plastic is rigidly attached to wood, metal, glass or ceramic—all of them relatively stable materi-als—the plastic may split, crack or bow between the fastening points. To avoid this problem, make the holes in the plastic larger than the fasteners, allowing the plastic to expand or contract in increments over its entire surface.

The slower expansion rates of the metal fasteners themselves can cause the expanding plastic to split or crack as it presses against the heads of the fasteners. You can prevent this by simply seating the nuts and bolts or rivets less tightly against the plastic.

Permanent Connections with Rivets and Washers

Quick fastening with pop rivets. Align the holes in the two pieces to be fastened, tapping them together if necessary. Insert the long mandrel of the rivet into the nosepiece of the pop-rivet tool, then insert the shaft of the rivet into the hole in the assembly. Place a back-up washer over the protruding shaft, and squeeze the handles of the pop-rivet tool together. Continue alternately squeezing and releasing the handles until the rivet is firmly seated and you feel increased resistance. Then squeeze the handles a final time to snap off the mandrel, leaving only the rivet head showing. If a nub of the mandrel protrudes, use a metal file to remove it.

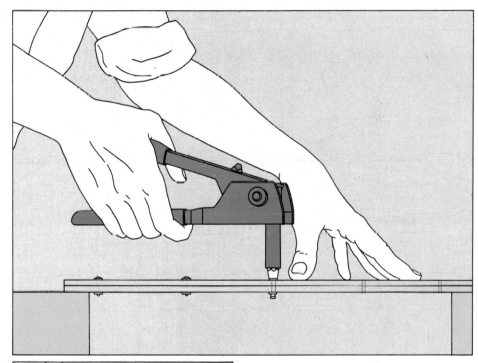

Installing a split rivet. Align the holes in the pieces to be joined, and insert a split rivet through them. Position the assembly, rivet head down, against a solid, padded surface, and slip a back-up washer over the split end of the rivet. With a screwdriver, bend each side of the rivet over the washer. Then use needle-nose pliers to curve the rivet ends back against the washer until the split rivet is snugly seated.

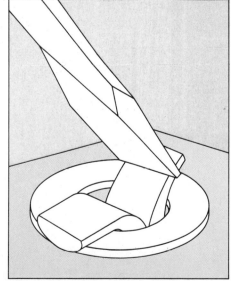

Hinging a Cover with a Pair of Pins

Making a pin hinge. Clamp or tape the movable piece into its closed position, and drill holes through the stationary pieces into both ends of the movable piece (*below, left*). Position the holes along the same axis at each end of the assembly, and keep the drill bit aligned with that axis as you cut the two holes.

Remove the movable piece and slightly enlarge its two holes. Then reassemble the pieces and drive long, smooth wire brads into the holes (*below, right*), using brads that fit snugly into the stationary piece. Check the assembly for free movement; file and sand away any plastic that interferes with the desired swing.

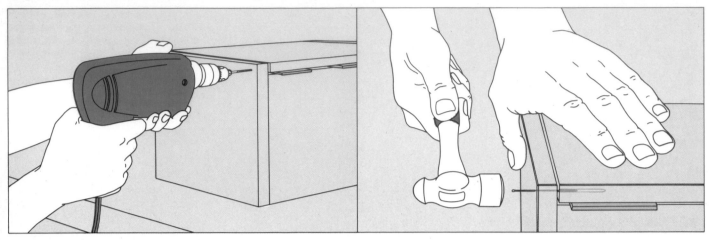

Thermoplastic Threads Formed by a Heated Screw

Shaping threads for a wood screw. Grip a bright, clean wood screw firmly in a pair of pliers and hold it over the flame of a propane torch (*right, top*) or a kitchen stove until it is deep blue. Then push it into a predrilled hole in the thermoplastic (*right, bottom*); the hole should be slightly smaller in diameter than the outside thread of the screw, and as deep as the length of the screw. When the plastic has melted and cooled into the shape of the screw, remove the screw with a screwdriver. In the final assembly (*inset*), use new screws of the same size.

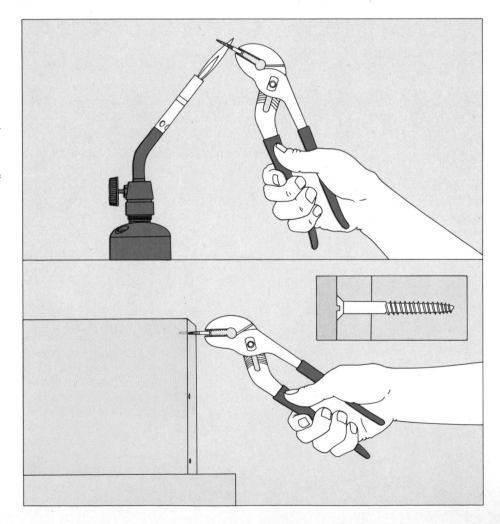

A Nut Sealed in Place for Extra Strength

1 **Installing the nut and bolt.** Drill a pilot hole for the bolt through the two pieces of plastic to be joined, making it the same size as the outside thread diameter of the bolt. Counterbore the hole in what will be the finished surface, making it large enough for a washer and a nut with ⅛ inch of clearance. Drop in a washer, insert a bolt from below and thread a nut onto the bolt. Using a socket wrench to hold the nut, tighten the bolt until the pieces are firmly connected.

2 **Embedding the nut.** Using an oil can with a small spout, cover the threaded bolt end with a drop of light machine oil; insofar as possible, keep the oil off the nut. Mix a small batch of plastic patching compound (page 105) to match the color of the plastic surface and pour it into the counterbored hole, filling the hole completely. Allow the patching compound to cure completely before removing the bolt from the nut, now hidden in the patching compound.

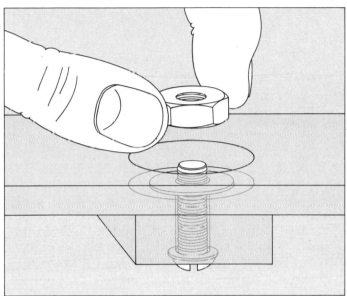

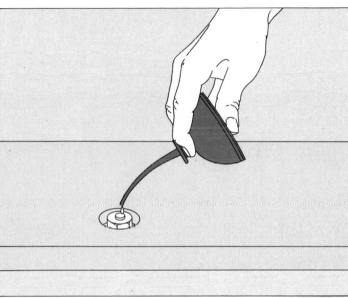

Threading a Hole with a Tap Wrench

Tapping threads in plastic. Mount a tap the size of the fastener in a tap wrench; dip the cutting threads in soapy water and, working clockwise, slowly begin to twist the tap into a predrilled hole. Keep the tap exactly aligned with the hole, and back it out frequently to clear it of plastic chips.

When you drill the pilot hole, make its length equal to that of the chosen fastener and its diameter halfway between the inside and outside diameters of the fastener's threads (inset).

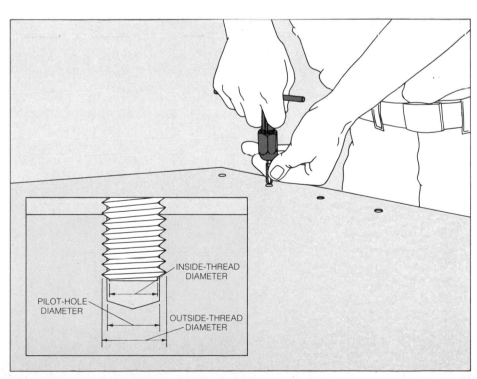

INSIDE-THREAD DIAMETER

PILOT-HOLE DIAMETER

OUTSIDE-THREAD DIAMETER

Parts Bonded with Cements, Glues or Solvents

Bonding is to plastic as nailing is to wood and welding is to metal: It is the method of choice for assembling or repairing. Compared to screws, bolts or nails, bonding has the advantage of distributing the load it bears over the entire length of the joint. Compared to welding, it requires far simpler equipment—often no more than a squeeze bottle for applying the adhesive and a jig to hold the bonded joint until it dries.

Plastics can be bonded with a wide variety of cements and glues, some of them familiar household products, others more esoteric (chart, opposite). The type of plastic dictates the choice of adhesive, and for most plastics there is more than one choice. Two common plastics, however, polyethylene and polypropylene, resist cements of all kinds—because of the waxiness of their surfaces, they are best joined with heat.

The preferred method for bonding most of the hard, glasslike thermoplastics such as acrylic, polycarbonate, polystyrene and the cellulosics is solvent cementing, a method peculiar to plastics. This technique takes advantage of the fact that these glassy plastics soften on contact with strong solvents. In the most common solvent-cementing process, the joint is assembled and a few drops of solvent—usually methylene dichloride, either pure or in combination with other substances—are applied to the seam with a brush or a squeeze-bottle dispenser. Capillary action draws the solvent into the joint, where it softens the abutting edges. These then intermingle and, as the solvent evaporates, create a bond that is integral with the plastic.

In a variation of this technique, only one edge of the abutting plastics is softened, but it is softened to a greater depth, producing a slightly stronger joint. The edge is soaked in a pan containing a shallow layer of solvent. When the object is assembled, solvent from the soaked edge softens the adjacent surface; the edges of the two pieces meld and harden into a strong bond.

Although solvent cementing, with its crystal-clear joint, is most appropriate for joining furniture and kitchen or bathroom fixtures made of glasslike plastics, it is equally effective as a joining technique for other solvent-sensitive plastics, including polyvinyl chloride (PVC). The only precondition for its use is that the plastic parts must fit together without any gaps—since the amount of plastic softened is not enough to fill any unevenness. Square and sand the edges as described on pages 37-38, but do not polish them. If you are working with transparent plastic, a handy check for the precision of the joint is the water test shown opposite.

Acrylic can also be joined with a two-part acrylic cement. Useful for rough-trimmed edges and along joints that could benefit from the reinforcement of an extra fillet of material, it consists of a combination of acrylic resin and a hardener, mixed to a syrupy consistency just before use. Unlike solvent cement, acrylic cement actually requires a gapy joint rather than a close-fitting one; the opening is often produced by beveling one of the edges to create a triangular crevice. To inject cement into the crevice, use a large syringe made of polyethylene, available in hardware stores and pharmacies.

On plastics that cannot be bonded with either of these two special cements, many familiar adhesives work well. Objects cast from thermosetting resins can be bonded with epoxy cements; contact cements will fasten polyurethane foam; ordinary white glue is suitable for joining polystyrene foam.

Other common adhesives can be used on solvent-sensitive plastics when the neatness and clarity of a solvent-cemented joint are not crucial. Household cement will bond cellulosics; model glue joins polystyrene. These viscous plastic cements contain quantities of the specific plastic on which they can be used. Similar adhesives are available for PVC and for acrylic; they further extend the range of ways in which those materials can be joined.

When plastics are fastened to other materials, such as wood or metal, the range of possible adhesives broadens. Many such marriages require a pliant bond, to allow for the differing rates at which the two materials expand and shrink with temperature changes. Flexible epoxies—those without fillers—and adhesives based on silicone or synthetic rubber work best for such joints. White glue and viscous plastic cements join some plastics to porous materials, such as wood, and the so-called super glues lock plastics to a number of other materials.

Whatever bonding method you use, always observe certain preliminaries. First, read the label on the adhesive container. Not only will the label verify that you have selected the best product for the job, but it will also provide storage and use instructions and alert you to curing times—how long the joint must be immobilized before it is ready for use.

Make sure, too, that the surfaces to be bonded are clean—free of grease, polishing paste and traces of adhesive from the masking paper. To clean glasslike thermoplastics, wipe with lint-free cloth or chamois moistened in isopropyl alcohol. Objects cast from thermosetting resins, such as epoxy and polyester, must often be sanded to remove mold-release compound and to roughen their surfaces for better adhesion. Finally, design the joint and the clamping setup before you begin to apply the cement. Most adhesives give you very little time to reconsider, and firm, steady clamping is essential in making a strong joint.

Because the fumes of plastic solvents, cements, and glues are often flammable and nearly always irritating, safety guidelines similar to those for liquid plastics (page 8) must be followed assiduously. Always work in a well-ventilated area. Take large jobs outdoors—but wait for warm weather; many cements cure slowly and form a weak joint at low temperatures. Do not leave cement and solvent containers open any longer than necessary, and avoid extended skin contact with any cement or solvent. Never use these substances after drinking alcoholic beverages (the fumes may combine dangerously with the alcohol), near an open flame, or while smoking.

A Guide to Choosing the Right Adhesive

Plastic	Bonded to: Itself	Reinforced plastics	Wood	Metal
ABS	Solvent	Synthetic-rubber adhesive	Epoxy cement	Epoxy cement
Acrylic	Solvent, thickened acrylic cement, two-part acrylic cement	Two-part acrylic cement	Two-part acrylic cement	Contact cement
Cellulosics	Solvent, household cement	Synthetic-rubber adhesive	Household cement	Contact cement
Polycarbonate	Solvent, thickened polycarbonate cement	Synthetic-rubber adhesive	Epoxy cement	Epoxy cement
Polystyrene	Solvent, model glue	Epoxy cement	Model glue	Super glue
Polystyrene foam	White glue	Contact cement	White glue, contact cement	Contact cement
Polyurethane foam	Synthetic-rubber adhesive	Epoxy cement, contact cement	Contact cement	Contact cement
PVC	Solvent, thickened PVC cement	Synthetic-rubber adhesive	Synthetic-rubber adhesive	Contact cement
Rigid thermosetting plastics (phenolics, epoxies, polyester, melamine)	Epoxy cement	Epoxy cement	Contact cement	Epoxy cement, synthetic-rubber adhesive

Fitting the adhesive to the job. In the chart above, the first column lists the plastics most commonly fastened with cements or glues. The other four columns list the preferred bonding agents for joining them to the materials at the top of the other columns. Three of these materials—reinforced plastics, wood and metal—account for the vast majority of surfaces to which small plastic parts are customarily joined in building or assembling larger objects.

If the chart lists more than one bonding agent, both are equally effective. In such a case, appearance or clamping requirements may affect the choice—a fast-setting adhesive is often preferable for a complex joint, which is likely to be difficult to clamp.

The bonding agents are identified by their generic names. Since they are sold in various forms and under various trade names, you will have to check the ingredients on the package label to be sure that you are getting the cement or glue you want. Most of these bonding agents are available at hardware stores, and household cement is of course available everywhere. For model glue you may have to go to a hobby shop. Special solvent and two-part cements are carried mainly by plastics suppliers, although solvent-cement kits for joining acrylic are sometimes sold at hardware stores as well.

A Water Test to Verify that Edges Fit Precisely

Locating surface irregularities. When you are joining clear plastics with solvent cement, a method requiring perfectly matched surfaces, dribble a few drops of water from a medicine dropper onto one face of the joint. Assemble the joint, then sight through the plastic to watch the behavior of the water. If it spreads in an even film throughout the joint, the joint is ready for cementing. But if the water puddles in some areas (inset), the edges require further smoothing. Continue to file, scrape and sand the unpuddled areas, repeating the water test from time to time. Be sure to dry before cementing.

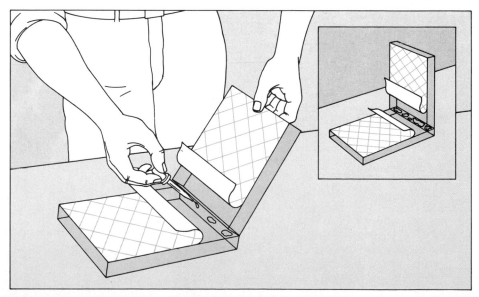

Solvent-softened Edges to Hold a Butt Joint

1 **Assembling and immobilizing the joint.** Butt together the pieces of plastic being joined, and anchor them at 3- or 4-inch intervals with lengths of masking tape stretched taut. Then carefully lift the taped plastic pieces and turn them over, propping them slightly above the work surface on wooden blocks spaced to give clearance to the seam. Brace the top with additional wood blocks to immobilize the joint and prevent the pieces of plastic from bulging upward because of the tension of the tape.

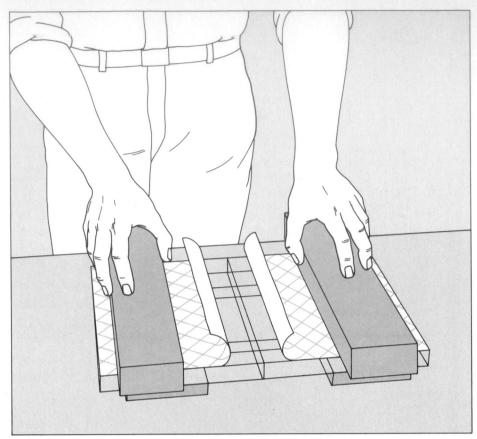

2 **Applying the solvent.** Dip a No. 1 artist's brush in the container of solvent, taking care to wet only the bristles so that the solvent does not attack the finish of the handle. Blot the brush on the lip of the container; then draw the brush along the joint, stopping to reload the brush if it becomes dry. Do not allow excess solvent to pool on the surface of the plastic. Capillary action will draw the solvent into the narrow aperture of the joint as you apply it. If the joint shows light-colored areas, an indication that the solvent has not penetrated the joint completely, make a second pass with the brush.

Let the joint stand undisturbed for at least 15 minutes. Wait at least four hours before putting any weight on the joint and before sawing, drilling or sanding in its vicinity.

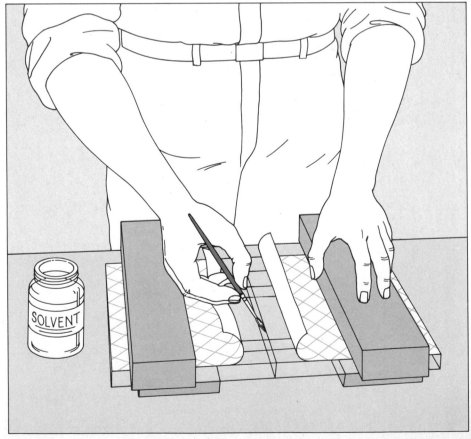

Using a Jig for Cementing a Corner Joint

1 Aligning and anchoring the joint. Make a right-angled jig from two pieces of ¾-inch lumber, one of which has been rabbeted to form a channel at the inside apex of the angle; nail the jig together. Position the plastic pieces in the jig so that the seam falls over the channel; this will prevent excess solvent from pooling between the wood and the plastic, blemishing the plastic. When the joint is properly aligned, secure the plastic with masking tape, as in this example, or with spring clips or large rubber bands.

2 Filling the joint with solvent. Fill the applicator bottle of a solvent-cementing kit with solvent; touch the needle-like tip of the bottle to one end of the joint and, squeezing the bottle gently, draw it along the length of the seam, dispensing an even flow of solvent. If capillary action fails to draw enough solvent into the seam to fill it, make a second pass, adding another bead of solvent. Wait at least 15 minutes before removing the plastic from the jig, and at least four hours before subjecting the bonded joint to any stress.

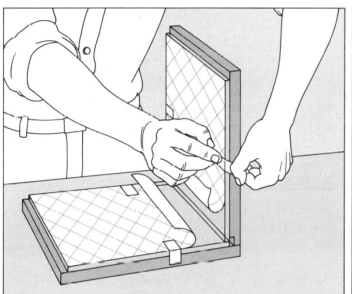

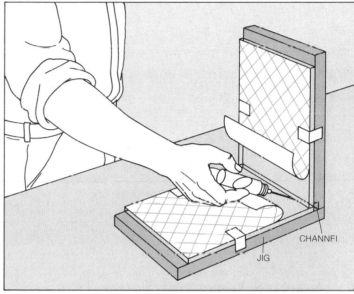

A Solvent Dip to Soften Parts for Joining

1 Softening one edge in solvent. Working in a well-ventilated area, prepare the dipping bath by arranging a pattern of paper-clip supports in the bottom of an aluminum pie pan or a glass baking dish; the pattern of paper clips should match the outline of the edge being softened—in this example, a plastic cylinder. Pour just enough solvent into the pan to cover the clips, and set the plastic on top of them; only the bottom edge of the plastic should touch the solvent. Leave the plastic in the solvent for six to eight minutes, so that the edge can soften.

To slow the evaporation of dangerous vapors during the dipping operation, use the smallest pan possible. For extra protection, you can fashion a cover for the pan from polyethylene film. Cut a hole in the cover just large enough to accommodate the plastic being softened, and tape the cover to the sides of the pan.

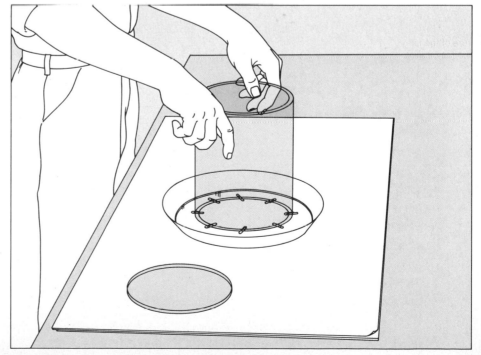

2 **Assembling the joint.** Remove the film and lift the plastic from the dipping bath, tilting the plastic so that any remaining solvent collects at one point on the edge, then blotting away the excess. Align the two parts being joined—in this example, the cylinder joins a circular base. Allow the assembly to sit undisturbed for a minute or two, while the solvent spreads from one piece of plastic to the other. Then weight or clamp the assembly for an hour; do not put any stress on it for 24 hours.

As soon as you remove the plastic from the dipping bath, cover the bath with a cookie sheet or a piece of polyethylene film, to slow evaporation of the solvent. The bath can be reused for additional joints or discarded.

Mistakes to Avoid in Solvent-cemented Joints

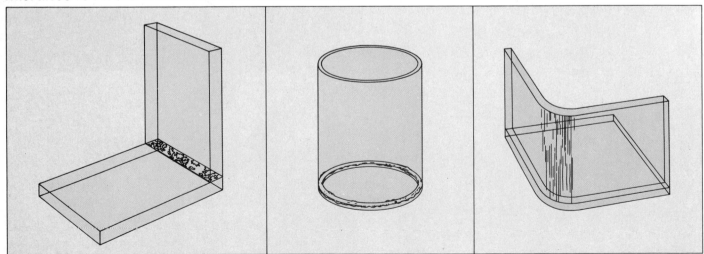

Three failed joints. A poor fit between the pieces in the joint at left caused an incomplete bond; because the solvent could not spread evenly throughout the joint, the seam is weak and unattractively blotchy. To avoid this problem, always be sure to scrape away any saw marks and sand the edges carefully before assembling any solvent-cemented joint.

The bulging joint at center was caused by premature clamping. The solvent in the dipped edge did not have time to soften the adjacent plastic; the pressure of the clamp against the still-hard plastic squeezed out the tacky material.

The fine cracks that mar the curved plastic surface at right were produced when solvent was

inadvertently spattered on the curved surface. The curved plastic, already stressed by heat-bending, was further weakened by the solvent, and the accumulated strains caused the web of cracks, called crazing. To reduce the chances of crazing, allow only the edge of the plastic to touch the dipping bath, and remove as little as possible of any protective masking paper.

Strong Acrylic Joints with a Two-Part Cement

Mixing the resin and hardener. Allow both the resin and hardener to reach room temperature, then pour the resin into a disposable glass or polyethylene container and add the prescribed amount of hardener, following the proportions listed on the label. Stir the mixture gently but thoroughly with a wooden tongue depressor for a minute or two, taking care not to introduce bubbles. If air bubbles do appear, let the mixture stand for three to five minutes, until the bubbles rise to the surface and are dispersed.

Working quickly—the syrupy mixture will harden to uselessness in about 25 minutes—fill the disposable syringe applicator by dipping the tip in the mixture and slowly pulling up the plunger. When you have finished with the cement, allow the leftover mixture to harden before discarding both the container and the syringe.

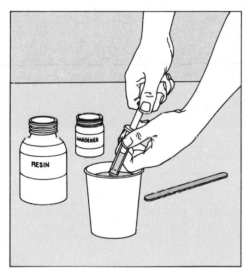

Acrylic-cementing a butt joint. Bevel the edge of one piece of plastic to an angle of 15°, as described on page 24, then butt the two pieces of plastic together, beveled edge up in the joint. Weight the plastic between wood blocks as shown in Steps 1 and 2, page 50. Leave a 1/16-inch gap between the two plastic edges and secure the underside of the gap with a temporary masking-tape seal.

To construct the seal, cut a strip of 1-inch masking tape slightly longer than the joint, and a strip of 1/2-inch masking tape, also slightly longer than the joint. Place the 1/2-inch strip down the center of the 1-inch strip, sticky sides together (inset), and attach the seal to the plastic. Lap the tape over the ends of the joint to form dams for the cement. The sticky margins will seal the joint, and the center strip will prevent the adhesive on the tape from interfering with the curing of the cement.

Beginning at one end of the joint, fill the joint groove with cement by depressing the plunger of the syringe and moving it back and forth along the joint; overfill the groove slightly, to allow for shrinkage as the cement cures. Let the joint stand undisturbed for at least four hours. After 24 hours, sand the bead of cement level with the plastic surface (pages 37-39).

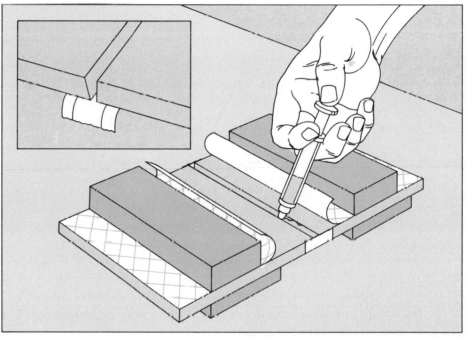

Acrylic-cementing a right-angled joint. Make a right-angled wooden jig using the method described in Step 1, page 51, but trim the outside edge of the jig's joint at a diagonal, forming a chamfered corner. Bevel the edge of one piece of plastic to an angle of 15°, and tape or clamp the two pieces of plastic against the outside of the jig. Position the unbeveled piece flat against the work surface and set the beveled piece against it, bevel facing out and 1/4 inch back from the edge of the unbeveled piece; this 1/4-inch ledge will later be removed.

Using the syringe, dispense a bead of cement into the grooved joint, overfilling it slightly to allow for shrinkage. While the cement is still viscous, shape it into a concave fillet with the tip of the syringe. Allow the joint to cure for four hours; then, for a stronger bond, make a second fillet on the inside of the joint. After 24 hours, use a table saw (page 24) and a medium mill-cut metal file (pages 37-39) to trim off the 1/4-inch ledge, leaving a smooth right angle.

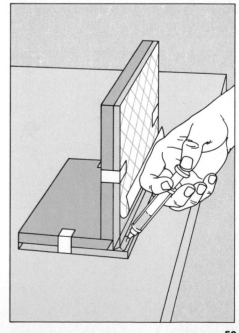

Heat as a Means of Merging Thermoplastics

Because of the low melting point of many thermoplastics, welding them with heat produces none of the showy pyrotechnics of metal welding. Nor does it require the same kind of heavy equipment. But like metal welding, joining plastic by means of heat produces a bond that is waterproof, airtight and, in some cases, fully as strong as the nonwelded areas of the material.

Virtually any thermoplastic can be welded. Thermosetting plastics, on the other hand, merely char or decompose if subjected to welding heat. The plastics most commonly welded are polyethylene, polypropylene and polyvinyl chloride (PVC). Polyethylene and polypropylene resist solvents and thus cannot be joined with solvent cement (*pages 50-52*)—a generally faster and neater technique. PVC pipe is joined with solvent cement, but heat welding produces stronger joints in PVC sheet and film. Nylon, which can be cemented only after special surface preparation, is often welded instead.

Like metal welding, plastic welding requires that the edges of parts be heated to a liquid and then held together until the molten material fuses and sets. Because the strength of a welded joint depends on molecular fusion—the uniform comingling and coalescence of the materials that are being joined—only identical thermoplastics can be successfully welded to each other.

The amount of heat needed to accomplish the necessary fusion—and the length of time that the plastic must be subjected to it—vary with the type and thickness of the plastic. Most can be sufficiently liquefied in less than half a minute at temperatures ranging from 430 to 560° F. Because these temperatures are relatively low, the heat for plastics welding can be generated by almost any heating tool or appliance—including a household iron or a kitchen range. However, welding is also done with certain specialized tools, such as hot-gas welding guns (*pages 58-63*).

The handiest tool for heating the ends of pipes, rods and small sheets of plastic is a hot plate—preferably one with a thermostatic control and a smooth aluminum or steel top. To keep the plastic from sticking, coat the top of the hot plate evenly with a thin layer of permanent nonstick spray, available at hardware stores. Never use wax, oil or silicone; they can be picked up by the plastic and will fatally weaken any weld. For the same reason, be sure to clean any excess melted plastic off the hot plate after each use, either by carefully wiping it with a rag while it is still hot or by scraping it off with a knife or a razor blade after it has cooled.

To melt the ends of small parts, heat can be most precisely applied with a hot scrap of metal. Commonly, a small piece of aluminum or steel sheet is heated on a range burner or a hot plate, or with a propane torch, and then touched briefly to the parts to be melted. Of course, the open flame itself must never contact the plastic—the material would ignite rather than melt. For the flexible seams needed to join thin plastic sheets or films, an ordinary household iron performs admirably, although a special tool such as a heating wand can speed up the job considerably (*page 66*).

No matter which heating method you use, plastic edges to be joined must be scrupulously clean and must be cut to fit each other precisely. Then, to maintain the precise fit, the joint should be aligned with a jig as it cools.

A simple two-part jig for flat or round objects can be made from two pairs of plywood strips, each pair nailed together at a 90° angle to form an L-shaped support (*Step 2, opposite*). In use, the two parts are positioned about an inch apart and the plastic sections being joined are seated in them, with the seam line in the open space. The jig aligns the joint and also supports the plastic sections while the weld is cooling.

A more complex two-part jig for precise alignment of small rods and tubing is mounted on the jaws of a vise; as the jaws of this jig are closed, the cylindrical pieces of plastic are moved smoothly together (*page 57*).

Because of the low temperatures involved, most plastics welding can be safely done in any room of the house. The worktable can be protected with a sheet of cardboard or plywood. Good ventilation, however, is essential. Any plastic, when melted, gives off noxious fumes, and the fumes of PVC can be deadly. If you are welding PVC, use a fan or two to disperse the fumes; avoid leaning close to the work, where the vapors are most concentrated.

The Handy Hot Plate for Minor Welding

1 Melting the edges. Set a thermostatically controlled hot plate to a temperature of about 500°. Hold the pieces that you are welding against the plate, at right angles to its surface, without applying pressure. When the lower edges of the plastic become hot enough that they are clear and pliable about ⅛ inch up from the surface of the hot plate, the thin layer that is in direct contact with the plate will be liquid and ready for joining.

2 **Joining the pieces.** Lift the two pieces of plastic straight up from the hot plate and place them in the two-part jig, butting them against the jig's L-shaped back *(below, left)*. Immediately slide them toward each other until their molten surfaces touch. Press them together just enough to form a small bead of plastic along the joint

(inset). Hold the pieces in position until the plastic cools. Then smooth off the bead with 280- to 400-grit wet abrasive paper or a sharp knife.

When fusing pipes and rods *(below, right)*, hold them in alignment by pushing them into the V formed by the bottom and the back of the jig.

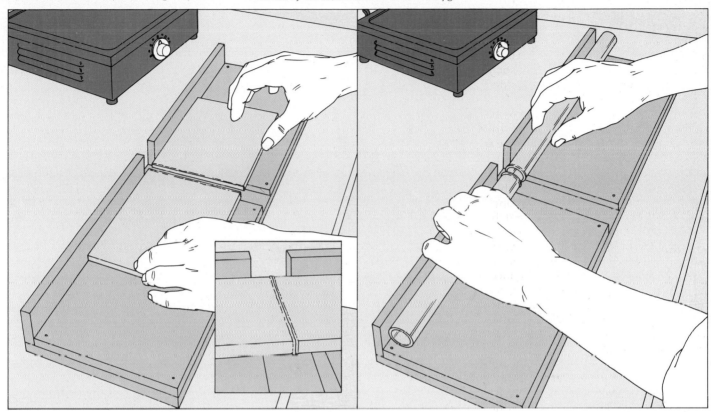

A Double-faced Heater Made of Scrap Metal

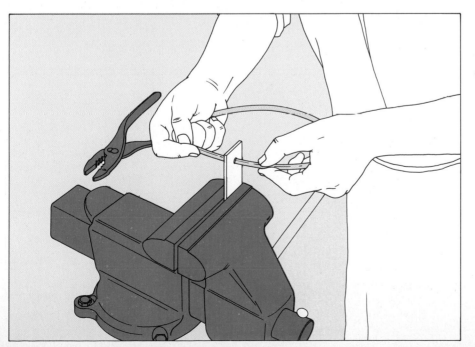

1 **Melting the plastic.** Heat a small scrap of aluminum or steel on a hot plate or over a gas flame. Using pliers, lift the hot metal and clamp it between the jaws of a vise. Touch the two pieces of plastic to opposite sides of the metal, holding them there until they soften and become clear to a depth of approximately 1/8 inch.

2 **Fusing the parts.** Pull the melted ends of the plastic straight out from the heated metal—without letting them slide—and quickly press them together firmly enough to form a slight bead at the joint. Hold the plastic until it cools and hardens. Remove the bead as in Step 2, page 55.

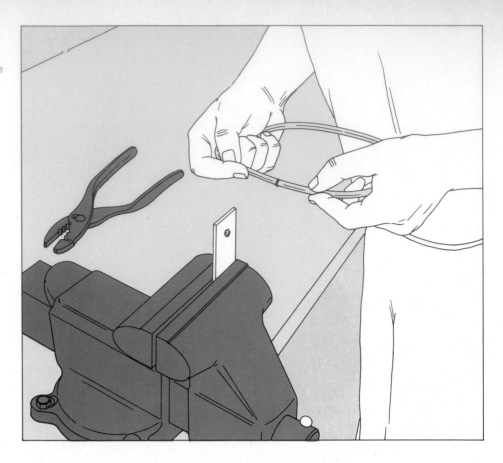

A Jig Designed to Align Parts Being Welded

1 **Constructing the jig.** Cut two pieces of 1-by-2 wood the same length as the jaws of the vise. Drill two small holes edgewise across each piece for a bolt-and-wing-nut assembly, positioning the holes about 1 inch from the ends of the wood; countersink the holes. Then drill a large hole through the face of each piece of wood, centering it exactly; make the diameter of this hole the same size as, or slightly smaller than, the diameter of the plastic tubing or rod being joined.

Cut a saw kerf, ⅜ inch deep and ¼ inch in from the edge, into each end of the wood to provide a slot for a hose clamp. Then cut each piece of wood in half lengthwise, dividing the large hole into a pair of semicircles.

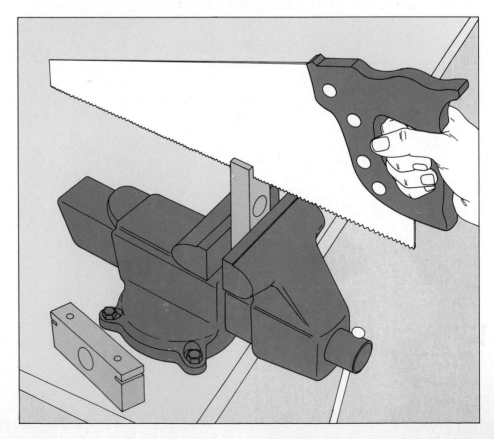

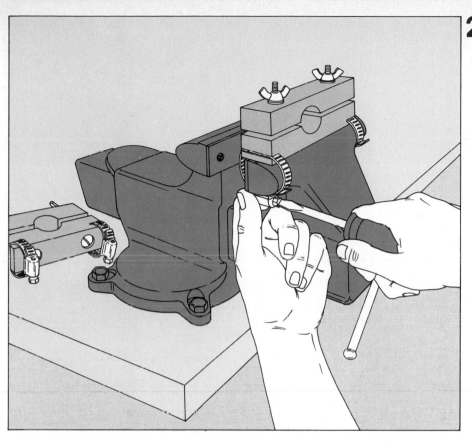

2 **Mounting the jig.** Insert bolts through the boltholes in the upper and lower halves of the jig sections, securing them with wing nuts. Loop a hose clamp through each saw kerf, and attach the two jig sections to the two jaws of the vise. Then loosen the wing nuts, and clamp a length of plastic tubing or rod in each of the two jig sections. Finally, loosen the clamps and adjust the positions of the jig parts until the ends of the plastic are perfectly aligned; tighten the clamps.

Joining Rods or Tubes in a Jig-equipped Vise

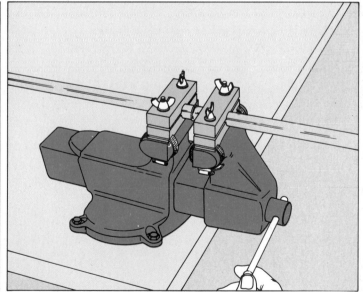

1 **Melting the jig-held plastic.** Open the jaws of the vise until the ends of the tubing or rod are about ½ inch apart. Then insert a piece of scrap metal, heated as in Step 1, page 55, between the ends of the plastic, and tighten the vise until the ends rest against the hot metal. Leave the metal in place until the plastic is soft and clear to a depth of about ⅛ inch.

2 **Fusing the ends of the plastic.** Open the vise just enough to free the softened ends of the plastic from the hot metal. Remove the metal and immediately close the vise, bringing the ends of the plastic together; apply just enough pressure to form a small bead of plastic at the joint. Let the plastic cool and harden, then remove it. Smooth off the bead as in Step 2, page 55.

Fast Welding with Heated Gas

A hot-gas welder, originally designed for such industrial uses as repairing plastic ductwork and storage tanks, also makes short work of household repairs, joining the broken parts of a plastic fan blade, patching a hole in a plastic garbage can, and restoring similar objects to active life. Welding can also be used to make fluid-proof liners for planters or custom-shaped storage tanks for household or hobby chemicals. The welder directs a jet of hot air or gas over a plastic filler rod, which softens and flows into the joint between the parts or around the edge of the patch, fusing the pieces together. A small home welder is not expensive, but you can often rent one from a tool-rental agency or plastics supplier.

The business end of a hot-gas welder consists of a nozzle-like handpiece containing a powerful stainless-steel heating element. Pressurized air or gas enters the handpiece through a hose and is warmed as it passes the element. It exits through one of several interchangeable tips that spread or concentrate the stream of air or gas, as the particular weld requires.

For most plastics welding, compressed air is satisfactory. It may be supplied by the type of compressor used for home spray painting—one that can deliver air at a rate of 3 cubic feet per minute under at least 15 pounds per square inch (psi) of pressure. Or, if a compressor setup is impractical, you can substitute bottles or tanks of pressurized air.

For polyethylene, use pressurized nitrogen gas, available in tanks or bottles. Air's oxygen would react chemically with heated polyethylene, breaking it down and weakening the weld.

Whatever the source of air or gas, the pressure will need to be reduced to working level, between 1½ and 3 psi, by a valve in the connecting hose. Adjusting the pressure also changes the temperature of the air or gas exiting from the tip. At low pressures, the air or gas passes the heating element slowly and may be heated to 600° F.; at higher pressures, the gas blows past the element quickly, in some instances reaching only 400°.

The right temperature for most plastics welding is between these extremes. Try welding a scrap of the same kind of plastic as what you plan to join. Check to see if it is properly heated (page 60, top), then adjust the gas flow. After welding, switch off the heat and let the gas flow continue for a few minutes to cool the stainless-steel heating element and keep it from burning out prematurely.

Although there are many ways to make a welded plastic joint, good preparation of the plastic and a good welding technique are essential. Clean the edges of the plastic thoroughly and roughen them with sandpaper. To start the weld, position the pieces carefully and, if necessary, hold them together with a push stick or clamps; then tack the joint in a few spots with concentrated blasts of hot gas.

The filler rod for the weld must be the same type of plastic as that being welded. You can use a strip cut from a plastic sheet, but extruded round and triangular rods, available in rolls of several sizes and lengths, are more convenient. When melting the rod into the joint, hold the welding tip ¼ to ½ inch above the plastic, and move it back and forth to distribute the heat evenly. Heat the rod only enough to liquefy its surface; its core should remain stiff enough to be pushed into the joint. Depending on the thickness of the stock, several rods may have to be laid in to fill the joint.

A slightly different welding technique is used for speed welding, a variant process that is particularly useful for projects involving many joints. In speed welding, a special type of slotted welding tip dispenses heated filler rod or filler strip directly onto the joint. This eliminates the need to fan the tip over the filler in order to heat it evenly. Also, because it is in effect one-handed welding, freeing one hand to hold the plastic in position, the weld can often be made without preliminary clamping and tacking.

Whatever the welding mode employed, joining plastics with heat is generally a safe procedure. Except at the end of the welding tip, the temperatures involved are moderate; no special protective clothing or gear is required, and the work surface needs no special insulation. Do not, however, aim the tip at your skin at close range. Also, be sure that the welding area is well ventilated: Melting plastics release noxious fumes; those of PVC are particularly dangerous to inhale.

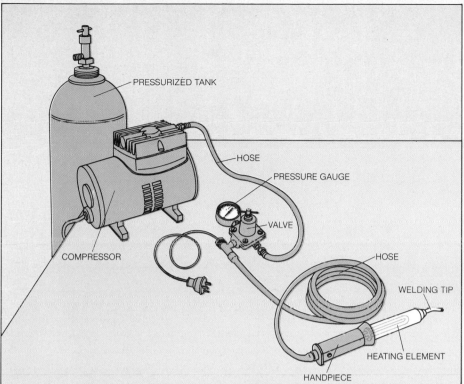

The welding apparatus. Air or gas from a compressor or a pressurized tank is pumped through a hose to a pressure gauge and valve. A second hose carries the air or gas from the valve to the handpiece; it is heated there, then exits through the welding tip. The heating element is heated by an electric wire that runs through the same hose as the air or gas. A second electric cord activates the compressor.

PRESSURIZED TANK

HOSE

PRESSURE GAUGE

VALVE

COMPRESSOR

HOSE

WELDING TIP

HEATING ELEMENT

HANDPIECE

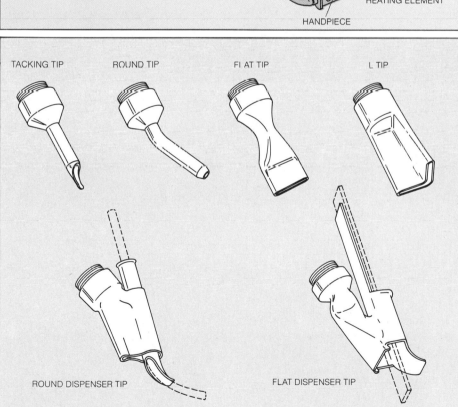

TACKING TIP

ROUND TIP

FLAT TIP

L TIP

ROUND DISPENSER TIP

FLAT DISPENSER TIP

BUTT WELD

SPEED-WELDED BUTT JOINT

OVERLAP WELD

T WELD

Four basic welds. For a butt weld—the commonest way to join two pieces of plastic end to end—the edges of the joint are beveled at a 30° angle, creating a 60° V groove. A ¹⁄₆₄-inch space at the bottom of the groove allows the molten plastic to penetrate the joint, and one or more filler rods are laid into the groove to fill it. When speed-welding a butt joint, you need not bevel the edges, but a ¹⁄₆₄-inch gap must be left between the pieces of plastic so that the filler strip penetrates.

For an overlap weld, often used in patching large holes, one piece of plastic is laid on top of another, and the right angle where they meet is filled with molten plastic. The work is then turned over and welded on the other side.

For a corner weld, layers of filler are similarly laid into the right angle formed by the two pieces of plastic. For a stronger corner weld, butt the two pieces of plastic together in a *T* and weld on both sides; the excess plastic on the outside of the *T* is then trimmed back to the edge of the melted filler rods reinforcing the joint.

A choice of welding tips. By shaping and directing the emission of air or gas in various ways, different tips increase the versatility of the hot-gas welder. The tacking tip, for spot welding a joint to hold it in place temporarily, has a sharply protruding end that focuses air or gas on a tiny spot. The all-purpose round tip, which is appropriate for most welding tasks, delivers a broader, more uniform flow than the tacking tip does. The flat and L tips are helpful for reaching into narrow spaces and tight angles. The round and flat dispenser tips, designed especially for speed welding, lay softened ribbons of filler rod or filler strip directly on the joint.

Judging weld quality. The three butt welds illustrated show the effect on a joint of differing amounts of heat. In the weld at top, near right, filler rods have not been heated enough to lose their shape and have fused incompletely. Correct this defect before adding any more filler rods; reheat the weld, and either move the welding tip more slowly along the joint or decrease the gas pressure so that the gas leaving the welding tip is hotter. The gently rounded top of the filler material in the weld at far right indicates that just enough heat was used to fuse the rods to each other and the sides of the joint, creating a strong bond. In the discolored bottom weld, either the tip was moved too slowly or the gas pressure was too low; the filler rods overheated and charred before they could fuse. This joint cannot be salvaged; cut away the burned material and replace it with new filler rods.

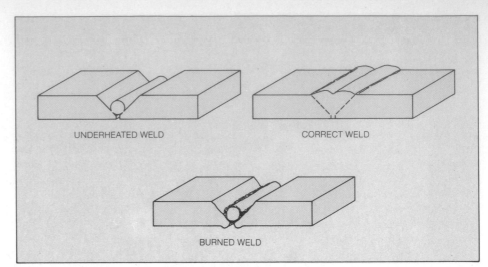

UNDERHEATED WELD   CORRECT WELD

BURNED WELD

Using a Filler Rod to Seal a Butt Joint

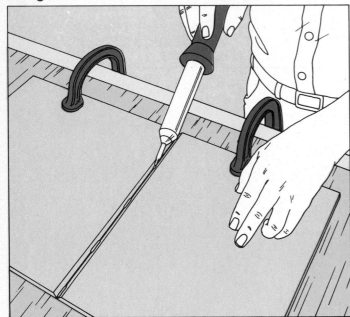

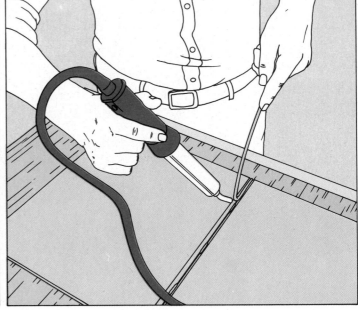

1 Setting up the weld. Place a piece of plywood on the work surface to help the plastic retain heat during welding. File or sand the two edges to be joined to 30° angles and butt them together, with a hairline gap at the bottom of the V; clamp the plastic to the plywood. Fit a tacking tip into the handpiece and turn on the heat and the gas or air; let the tip warm up for two minutes. Direct the heated tip at a single spot on the joint, holding the end of the tip lightly against the plastic. When the plastic softens, move the tip about 1 inch along the joint, fusing the spot. Repeat this tacking procedure at 3-inch intervals along the joint. After the last tacking has cooled for about one minute, remove the clamps.

2 Welding the joint. Replace the tacking tip with a round tip, using work gloves to remove the tacking tip if it is still hot. Allow the new tip to heat up; then, holding the handpiece in one hand and a filler rod in the other, set the rod in the groove at one end of the joint. Holding the tip ¼ to ½ inch above the rod, move the tip back and forth in a fanning motion across the joint, so that both the rod and the edges of the groove melt equally. As the rod melts, it will turn clear; when it reaches this stage, push it gently down into the groove. Move the handpiece along the joint, continuing to melt both the rod and the groove until the entire joint is filled.

If you run out of filler rod before the weld is complete, taper the end of a new rod and soften it with the welder. Lap the tapered end over the end of the rod in the groove and continue.

3 **Cutting off the rod.** Warm the filler rod about ¼ inch beyond the end of the joint, then bring the welding tip close to the rod and, without fanning the tip, apply a steady stream of hot air to one spot on the rod. Meanwhile, pull gently on the rod; it will thin and then separate.

Continue to lay filler rods into the groove until the surface is slightly higher than the surrounding plastic. When the welded joint is barely warm to the touch, in three to four minutes, cut or file off the ragged ends of the rods.

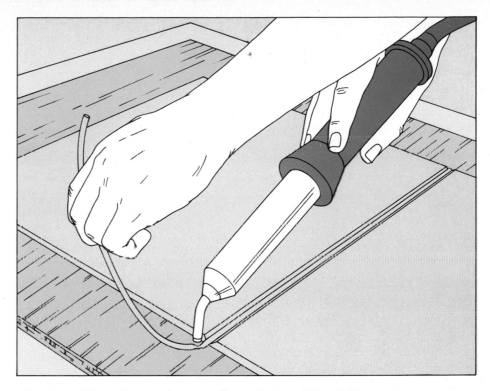

A Strong Angled Joint Made with a Wooden Jig

1 **Starting a corner T weld.** Construct a right-angled jig by nailing together two boards slightly smaller than the plastic sheets being joined. Clamp one sheet upright, against the outside of the jig, aligning the bottom edge with the jig bottom. Place this assembly on top of the other plastic sheet, ⅛ to ¼ inch from its edge, forming a T joint with one short arm. Using the jig as a brace, clamp this assembly to the work surface. Begin welding as in Steps 1 through 3, opposite and above, laying filler rods into the short arm of the T. Allow the weld to cool for three to four minutes and then remove the clamps.

2 **Completing the T weld.** Carefully reposition the work so that the jig rests against the outside joint, in the short arm of the T. Prop the jig up with a scrap of plastic to bring it even with the horizontal plastic sheet. Secure the new arrangement to the work surface with clamps. Weld the inside joint as in Step 1. When the weld has had time to cool, remove the clamps and the jig. Using a file or a saw, trim away the excess plastic on the short arm of the T, leaving only the fillet of plastic that forms the outside joint. Also cut or file away the ragged ends of the filler rod, as in Step 3, above.

Hole Patches, Big or Little

Plugging a small hole. Fit a round welding tip into the handpiece, and hold the end of a filler rod just outside the edge of the hole—in this case, a hole in an automobile windshield-washer reservoir made of polyethylene. Using a fanning motion, soften the filler rod and the plastic surface beneath it. As the two soften, press the rod over the hole. Cut off the excess filler rod as shown on page 61, Step 3.

For a slightly larger hole, lay the molten rod across the opening in several passes, side by side. Or cover the opening by working in a spiral pattern, from the edge of the hole in to the center.

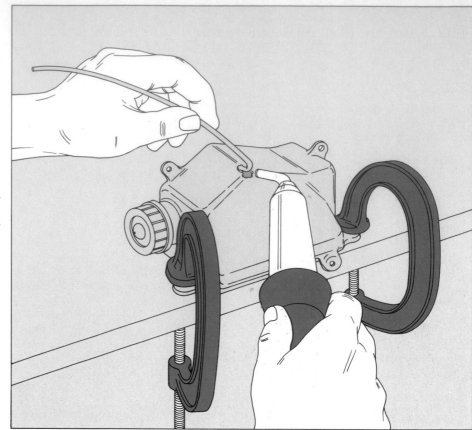

Patching a large hole. Cut a patch from plastic of the same type and thickness as the plastic in the object being repaired, making it at least ¼ inch larger than the size of the hole. Lay the patch over the hole and hold it in place temporarily with a wooden stick while you spot-weld the patch to the object (page 60, Step 1). Exchange the tacking tip, used in spot welding, for a standard round tip, and weld filler rod into the joint around the edge of the patch (page 60, Step 2). When you change tips, be sure to protect your hand from the tip's hot metal.

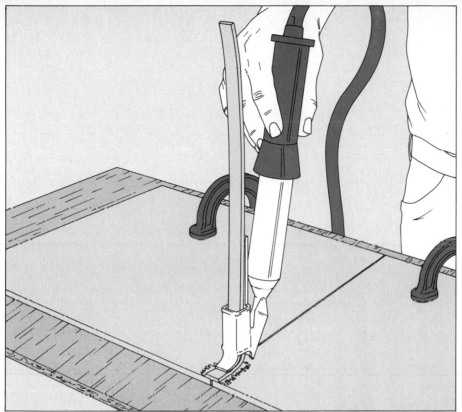

Using a Speed-welding Tip

1 **Tacking the joint.** Hold or clamp the plastic sheets in position, leaving a ¹⁄₆₄-inch gap between them. Then attach a high-speed tip to the welder; in this example the tip is designed for a flat filler strip, commonly used for the butt joint illustrated. Insert the strip into the slotted feeder and hold the heated welding tip almost upright, ⅛ to ¼ inch above one end of the joint. When both the tip of the filler strip and the plastic beneath it are melted, use the curved nose of the feeder to push the molten strip into the molten plastic, tacking the end of the joint.

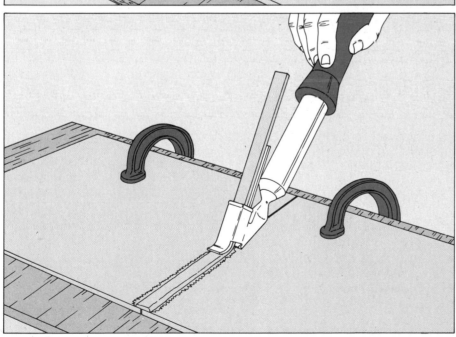

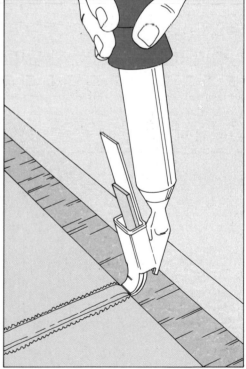

2 **Filling the joint.** Move the welding tip, still held upright, along the joint, feeding and pressing more filler onto the plastic until you have fused about an inch. Then drop the handpiece to about a 45° angle and pull the tool slowly along the joint, allowing the filler to feed out of the slotted welding tip in a continuous ribbon. As you work, continue to press together the molten filler and the molten plastic beneath it.

3 **Finishing the weld.** At the end of the joint, tilt the handpiece forward and dig the tip of the nose into the filler strip, severing it. Lift the tip back and away from the weld. Immediately remove the leftover filler strip from the hot tip, before it has a chance to harden and stick to the tool.

Hems or Seams in Flexible Film

The edges of plastic film and flexible plastic sheets can be heat-welded to form a variety of hems and seams that are useful for making such household items as shower curtains, garment bags and fitted seat covers. An ordinary household iron is surprisingly handy for these fabrications, efficiently fusing such diverse plastics as polyethylene, polystyrene and polyvinyl chloride. But for very fine seams, such as those that seal the edges of plastic-wrapped prints and posters, you may need to rent special sealing tools—bar sealers and heat wands—from a plastics supplier or a hobby shop.

Insulate the work surface from an iron's moderate heat with a sheet of heavy cardboard. To keep the plastic from sticking to the iron or the cardboard, sandwich it between two sheets of nonstick material such as cellophane or silicone-coated paper—most familiar as the peel-off paper on adhesive-backed plastics. If the iron has a nonstick coating, only the bottom layer of nonstick protective covering will be needed. Generally the iron should be set in the synthetics range and applied to the plastic for just a few seconds, but exact fusion times and temperatures depend on the composition and thickness of the plastic.

Experiment with scraps before attempting an actual project.

To guide the iron along a straight seam, butt it against a wooden straightedge; for a curved seam, cut a template from wood or heavy cardboard. No guide is necessary for seams made with heat wands or bar sealers, which will form straight seams automatically.

The precision of seams made with wands and bar sealers suits them well to a technique called shrink wrapping, which uses a special plastic. As the term implies, this material can be wrapped around an object, sealed at the edges, and then shrunk—usually with a heat gun—to conform exactly to the shape of the enclosed object. Shrinkable plastic can also be sealed with an iron, but the seams will be wider and less attractive.

Another sort of shrinkable plastic comes in tubular form and is most commonly used as a sleeve to gather wires into neat bundles. When heated, this tubular plastic shrinks to half its diameter without losing length. It is available in diameters that shrink to .023 to 1 inch—in any length—and can be shrunk with a hot-air gun, an industrial hot-air blower or even with the heat of a cigarette lighter, because the material is nonflammable.

A Choice of Three Hems and Two Seams

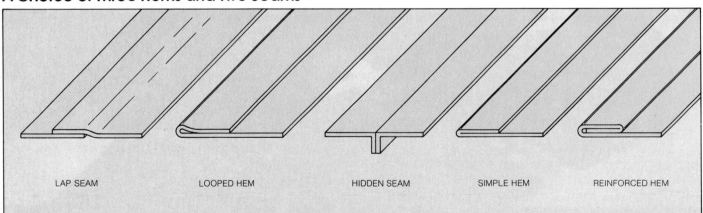

LAP SEAM   LOOPED HEM   HIDDEN SEAM   SIMPLE HEM   REINFORCED HEM

A range of seams and hems. Both the lap seam and the looped hem are made at the edge of the worktable, with the excess plastic overhanging the edge *(opposite, top);* the hidden seam and the other two hems are made at the center of the worktable, with the work supported on both sides *(opposite, bottom).* The lap seam joins two pieces of plastic film by simply overlapping and fusing their ends. The looped hem is folded back on itself and fused in a way that leaves an unsealed channel; a wire or a cord can be threaded through the channel. The hidden seam is formed by laying two sheets of plastic face to face and sealing their edges; when the sheets are opened out, the seam is on the underside of the work. The simple hem consists of a single fold, sealed on the underside of the work; for the reinforced hem, the fold is doubled.

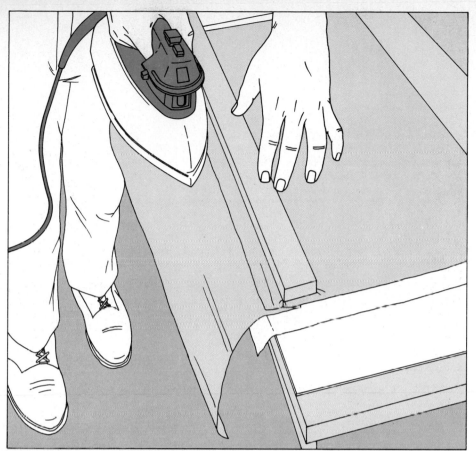

Forming a lap seam or a looped hem. Cover
the work surface all the way to its edge with a
nonstick protective covering. For a lap seam,
overlap the two pieces of plastic about ¾ inch and
position one edge of the seam along the edge
of the work surface. Weight the seam along its
center line with a straightedge. (If your iron
lacks a nonstick base, place a layer of nonstick
covering between the straightedge and the
plastic.) Beginning at one end of the seam, lower
the iron against the plastic, using the straightedge as a guide, and press down. Lift the iron,
move it along the edge and press again. Continue until the seam is fused along its entire
length. Then remove the straightedge and seal
the other half of the seam.

For a looped hem, fold the plastic back on itself
and line up the cut edge of the plastic with the
edge of the worktable. Place the straightedge
over the hem, covering enough of the fold line
to provide a channel of sufficient width for the reinforcing cord. Fuse the area between the
straightedge and the edge of the table.

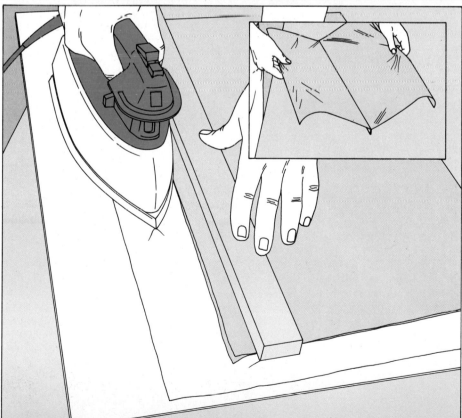

Fusing a hem or a hidden seam. For the
seam, spread a sheet of nonstick protective covering over the work surface and lay the two
sheets of plastic face to face over it, with their
edges aligned. If, unlike the iron shown here,
your iron does not have a coated soleplate, top the
plastic with a second layer of protective covering. Place a straightedge over the layered materials, about ¾ inch in from, and parallel to, the
edges of the plastic. Press down on the seam with
the iron, as described above, using the
straightedge as a guide. When the plastic sheets
have cooled, spread them open, turning the
seam to the back *(inset)*.

For a simple hem, fold the plastic back on itself
once; for a reinforced hem, fold it twice. Sandwich the plastic between two layers of protective
covering and place a straightedge along the
hem, covering about half its width.
Using the straightedge as a guide for the iron,
press down on the hem as for the seam above.
Then remove the straightedge and press down
the free edge of the hem in the same way.

Shrink-wrapping with Heat Wand and Blower

1 **Shrouding the object with film.** Set up the heat-wand assembly, including the roll of plastic film, the heat wand and its heatproof pad, and the hot-air blower used in shrinking the film. Turn on the switch at the control box to ready the wand; it will not actually heat, however, until you push the finger switch on the handle of the wand itself. Unroll enough film to cover the object being wrapped, and slip the object between the two layers of film. Slide the object toward you until it touches the seam at the end of the film.

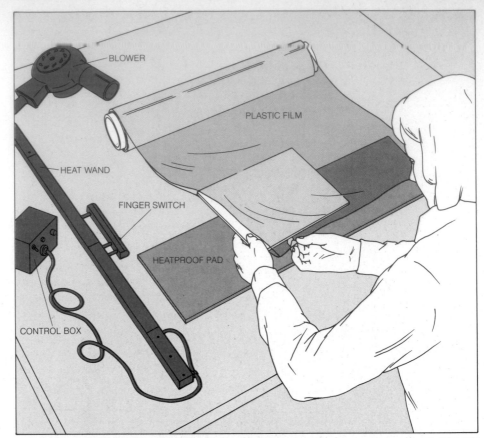

2 **Cutting and sealing the film.** Set the heat wand on the plastic about 1 inch in from the edge of the object, pressing the film against the heatproof pad. Depress the switch on the handle of the wand, activating the heating element. Gently tug at one end of the plastic film until it separates along the wand, generally after a second or two. Turn the plastic-covered object and seal the remaining sides, about ½ inch beyond its edges. Nick this airtight plastic envelope with a razor blade in some inconspicuous spot to let air escape when the film shrinks.

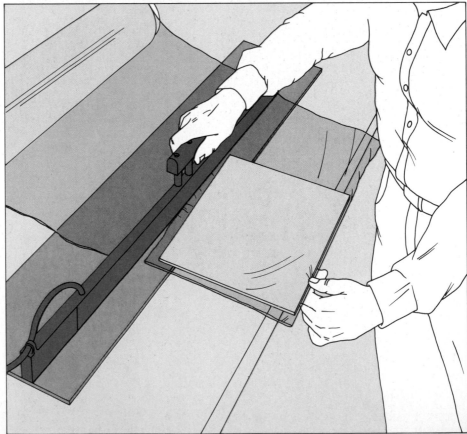

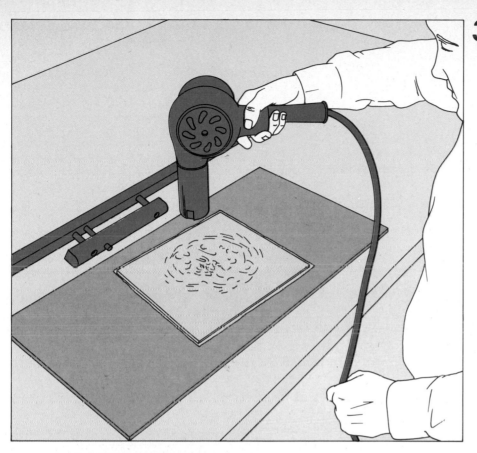

3 Shrinking the film. Rest the plastic-wrapped object on a protective sheet of cardboard and direct a stream of warm air against it with the hot-air blower, moving the blower in a circle and holding it about 3 inches above the surface of the plastic. When the plastic on this side begins to bubble and wrinkle, turn the work over; heat the other side until it bubbles, then pulls taut and smooth. Repeat to smooth the first side.

Shrink Tubing Used to Bundle Wires Together

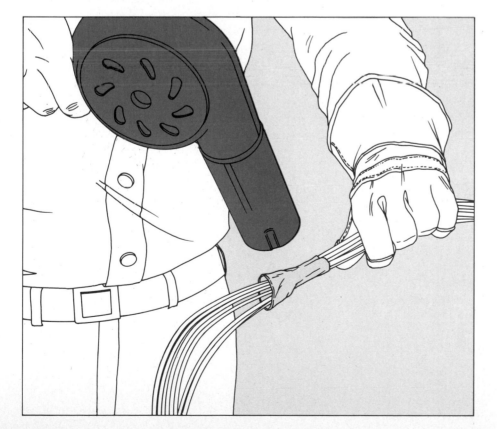

Gathering wires into a bundle. Estimate the diameter of the bundled wires and choose shrink tubing manufactured to shrink to that diameter or slightly smaller. Cut tubing to the desired length; it can cover the entire length of the wires or grasp them only at intervals. Slide the tubing over the wires. Sweep a hot-air blower back and forth over the tubing until it grips the bundle tightly. A similar heat-shrinkable tubing, thick enough to protect bare wires in low-voltage wiring splices, is available at electronics-supply stores; it must be heated with the open flame of a match or candle and thus should not be used to gather already-insulated wires, whose sheathings could be ignited.

A Primer on Poured Shapes

Line-for-line copy. A duplicate doorknob cast from liquid plastic resin has hardened and is ready to be removed from its synthetic-rubber mold. Made in two parts, the mold reproduces the contours of the original doorknob, or master. The mold, too, is a casting: The rubber compound is poured, in two stages, into a box containing the master. Once the compound hardens, the master is removed and the cavity becomes the mold for the new item. In lieu of the wooden box, built here of scrap lumber, the mold frame could be made of plasticine (*page 78*).

Although rigid plastics have many useful applications, it is the liquid plastics that truly capitalize on the material's major characteristic—its ability to be molded into any shape desired. Indeed, most of the plastic objects found in the home are formed by a variety of molding processes that turn liquefied plastic into everything from contoured tool handles to the appliance housings that neatly follow the functions of the working parts within.

In factories these processes are handled by sophisticated machines that heat and shape the plastic in one continuous operation. Phonograph records, for example, are made by compression molding: The halves of a 200-ton press melt the plastic and also form the grooves. Camera parts are formed in an injection-molding machine, which liquefies plastic pellets and squirts them into a mold under enormous pressure, as much as 40,000 pounds per square inch—all in a cycle that may consume no more than 10 seconds. And the familiar squeeze bottle is made by a blow-molding machine, which forces the liquid plastic against the walls of a mold in a repetitive process that is much like blowing up an endless succession of balloons.

None of these processes can be duplicated in the home workshop, but a surprising number of the objects they create can be produced by simpler means. Liquid resins can be cast in handmade molds—of clay, plastic or rubber—that have been shaped around a master, or model; the master may itself be handmade. Another technique is to layer liquid plastic and fabric over a homemade wooden form. Though neither approach requires machinery, they both call for patience, dexterity, and the ability to think in three dimensions. In short, modeling objects in plastic is very like creating a work of art.

Working Safely amidst the Vapors

Casting or laminating plastics releases toxic vapors that require more filtering than a paper dust mask can provide. The answer is a charcoal-cartridge respirator. This piece of equipment consists of a rubber face mask that can be fitted with a replaceable cartridge containing charcoal granules to filter out harmful vapors. On some models the respirator can also be fitted with a fabric filter for dust. Use of both dust filter and cartridge is strongly advised for sanding newly cast fiberglass laminate or for sawing finished laminate.

Respirators, filters and cartridge refills are available at large hardware stores, auto-paint stores and dealers listed under "Safety Equipment" in the classified directory. Some professionals recommend replacement of a cartridge after about 10 hours of use; certainly it should be replaced as soon as any smell or taste is detected. Dust filters should be replaced whenever breathing becomes difficult or after a full day of exposure to dust.

Fit is important and should be checked according to the manufacturer's instructions. Clean the mask after use, and store it in a sealed plastic bag.

Rigid Molds to Shape Liquid Plastic Resins

Many plastic objects found in the home—from doorknobs to bathtubs—began as liquid plastic resins that were shaped and hardened in molds. In factories these molds are made of hardened steel, to enable them to withstand the wear and tear of mass production. But at home, where the operation is on a much smaller scale, liquid plastics can be shaped in molds made of such commonplace, easily worked materials as plaster, wood or synthetic rubber. The latter is a molding compound available at plastics-supply houses.

Molds used for forming liquid plastics fall into two categories. Either they are cavities into which liquid epoxy or polyester is poured, or they are contoured forms over which successive layers of plastic are laid, usually in conjunction with a reinforcing material such as fiberglass. In a variation on this latter technique, shallow reliefs, such as ceiling medallions, are cast in surface molds.

The first step in casting anything in a hollow mold is to examine the object you will duplicate, called the master. Depending on its shape and surface details, you may choose either to make a rigid plaster mold *(opposite and pages 72-75)* or to use a flexible mold of synthetic rubber *(pages 78-83)*.

If the master has a complex shape, with reverse curves, undercuts and many incised details, a flexible rubber mold will be easier to pull cleanly from its surface. More durable than plaster, a synthetic rubber mold is also preferable if you plan to make many castings of the same shape. Under any other circumstances, however, a rigid plaster mold is better because it is so much cheaper. Make it of molding plaster, which hardens more slowly than plaster of paris, giving you more time to work. Molding plaster, commonly used for wall patching, is available at most building-supply stores.

In some instances, a plaster mold is poured in two parts in a frame built around the master; in other cases, it is made by applying plaster to the surface of a master in multiple sections. This choice, too, depends on the complexity of the master. To determine how many

mold parts you will need and where their separations will fall, you must analyze the shape of the master. As a general rule the mold will separate along the widest part of the master. But you must also look for reverse curves or undercuts that would interfere with the removal of the mold unless it subdivides into additional pieces. When you have located the dividing lines between mold parts—called the parting planes—draw them on the master with chalk or a marking pen.

The frame used to contain the liquid plaster for a two-part mold may be a plastic container or a cardboard box. To seal the seams of the box, use masking tape or the pliable rope caulking often used for sealing around windows. If you do not have a ready-made container of the right size, you can build the frame with scrap lumber or shape it with potter's clay *(page 79)*.

In preparation for pouring, you will also need to plan how to immobilize the master within the frame. If you have made the frame of clay, you can simply add a pedestal of clay for the master or use the technique shown on page 82, Step 1. A wooden, plastic or cardboard frame requires more ingenuity. Depending on the location of the parting planes, you may be able to suspend the master from a wooden crosspiece nailed to the top of the frame. Otherwise, anchor it to the side of the frame with a wood spacer, fastening the spacer to the master with existing hardware on the master, such as screws or bolts, or with glue; then attach the spacer to the frame with screws.

Whatever anchoring arrangement you use, keep in mind that the anchor also creates the sprue hole for the mold—the hole through which the casting material is poured.

Before the master is secured in the frame, it must be coated with a parting agent to prevent plaster from adhering to it. The best all-purpose parting agent is petroleum jelly, heated to the melting point in a double boiler, then evenly applied to the surface of the master with an artist's brush. For a final smoothing, the coat of jelly is then reheated with a heat gun or a hair dryer while any excess jelly

is blotted up with the brush. Another parting agent—one that is simpler to apply but leaves a less polished surface—is liquid dishwashing detergent. Also applied in a thin coat with a small brush, the detergent must dry for about five minutes before the plaster comes into contact with it.

Timing is important in mixing and using plaster to make a rigid mold. The prepared plaster will remain thin enough to pour for three to five minutes. After that, it will be about as thick as cake frosting and can be laid over the master by hand for up to another five minutes. Beyond five minutes, it will usually become cakey, suitable only for patching; in five minutes more, it will become too granular to use at all.

To mix plaster and water, use a flat-bottomed plastic mixing pan and put into it roughly four fifths as much water as the liquid plaster you will need. Sift plaster dust into the pan through your fingers, sprinkling it evenly over the surface of the water until it begins to form small islands of dry plaster on the surface. When these islands take about 10 seconds to become moist, blend the plaster solution with a large metal mixing spoon, stirring it for no more than 30 seconds. It is now ready for pouring.

Although the plaster mold will be hard enough to remove from the master in 15 to 30 minutes, it must dry, or cure, for five to seven days before use. Otherwise, the moisture remaining in the plaster might ruin the casting. When all moisture has evaporated from the plaster, the mold will feel much lighter and its surface powdery rather than clammy. You can accelerate the curing process by putting the mold in a 150° F. oven for 12 to 24 hours, depending on its size.

The last step in making the mold is to check the mold's interior surface for flaws. Use paint thinner to remove any petroleum jelly left by the master, then fill or rebuild any pockmarks or chipped edges with spackling compound. Smooth away any surface roughness with a fine grade of steel wool. If there are cracks or breaks in the cured mold, they can be repaired with white glue.

Handling Molding Plaster

☐ Never use stored plaster without first mixing a test batch; if it has absorbed moisture from the air, it will not harden properly.

☐ Control the rate at which liquid plaster thickens by altering water temperature, adding salt or vinegar, or adjusting stirring time. Warm water speeds hardening; small amounts of salt accelerate hardening, and up to a cup of vinegar retards it; lengthy stirring makes plaster harden faster.

☐ Cover the work surface with a sheet of polyethylene plastic; plaster does not adhere to this material.

☐ Simplify cleanup by coating the mixing pan and mixing spoon with a light film of kerosene before you start the mixing process.

☐ Rinse your tools in a bucket of water before plaster hardens on them. Let the plaster settle, then dispose of solids and liquids separately. Never pour plaster into your plumbing.

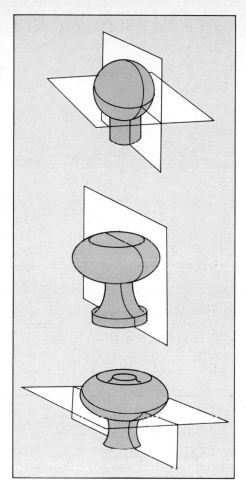

Suiting the Mold Type to the Shape of the Master

Finding the parting plane. The sections of a mold meet each other along a flat surface, called the parting plane, which is positioned so that each section can be removed from the master without interference. The number as well as the location of the planes is determined by the shape of the master—as illustrated by the three differently shaped doorknobs shown at left.

With its symmetrical shape, the top doorknob can have either a vertical or a horizontal parting plane—a two-part mold would separate cleanly in either direction. The undercut sides of the middle doorknob require a mold that separates vertically, to prevent interference from the base of the knob.

The bottom doorknob requires a mold with two parting planes. Because the modeling at the top of the master prevents mold sections from separating vertically, one parting plane cuts horizontally through the knob's widest part, sectioning off the top. However, the reverse curve in the base of the knob would prevent the removal of a single mold section there, so a second parting plane runs vertically through the knob, from its base to the horizontal parting plane.

A Two-Part Mold Made of Poured Plaster

1 Constructing a fallaway frame. Using scrap lumber, build a box with four sides and a base whose interior length, width, and depth are roughly 1½ inches greater than the dimensions of the master—in this example, a decorative coat hook. Connect the sides at the corners with wood screws, then nail the base to the sides with common nails or brads.

Draw the parting plane on the master with a marking pen. On the inside of the box, draw a cross mark at the point where the master will be anchored. Extend the horizontal line of the cross mark to use as a guideline in pouring the plaster. Then draw a matching cross mark on the outside of the box.

Seal the inside joints with masking tape, pressing strips of tape into the corners and over the seams between the bottom and the sides *(inset)*. Coat the interior of the box thoroughly with a thin layer of petroleum jelly.

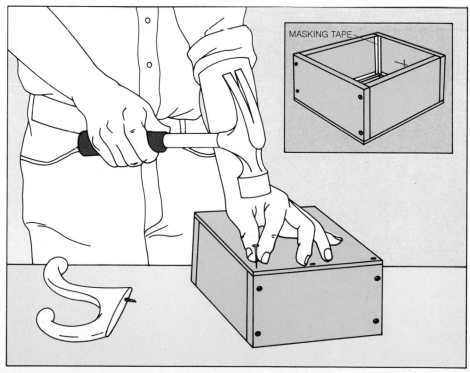

MASKING TAPE

2 **Anchoring the master in the frame.** Cut a wood spacer to fit the surface on the master to which it will be attached. Fasten the spacer to the master with glue or with any existing hardware—in this example, the coathook screw. Then coat both the master and the spacer with petroleum jelly *(page 70)*, taking care not to disturb the parting lines drawn on the master. Position the master in the frame, lining up the horizontal parting line with the cross-mark extension on the box, and drive a screw through the box into the spacer. To avoid marring the jelly-coated master, hold the work only by the spacer throughout this operation.

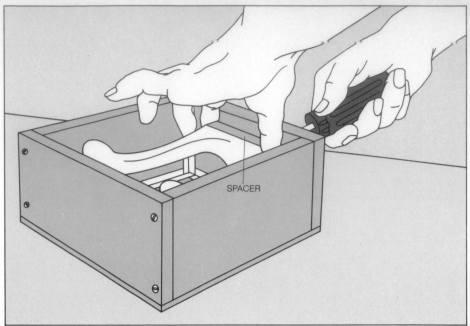

SPACER

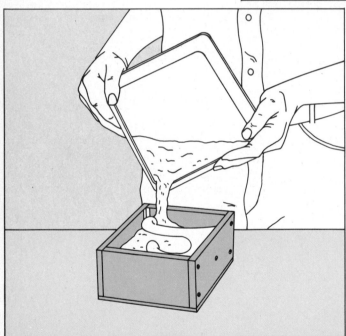

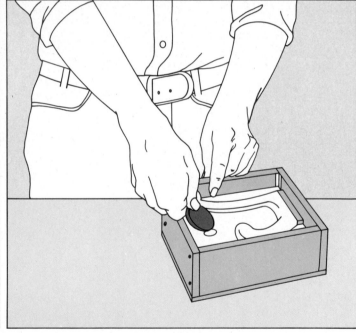

3 **Pouring the first mold section.** In a shallow pan, mix enough plaster to fill the frame up to the horizontal parting line, marked on the side of the box. Immediately pour it in a steady stream into the frame until it reaches the parting line. Tap the sides of the box several times with your hand to release any bubbles from the plaster.

4 **Keying the parts.** When the plaster is firm but not yet hard (in about 15 minutes), press the tip of a spoon into its surface to carve out a semi-circular indentation—called a key—that will be used in aligning the two parts of the completed mold. Turn the spoon 90° from the direction of the first key and make a second indentation, at right angles to the first.

When the plaster has set (usually in another 15 minutes), coat its surface with a thin layer of petroleum jelly. Mix a second batch of plaster and pour it into the frame, filling the frame completely. Again, gently tap the sides of the frame to remove any air bubbles.

5 **Dismantling the frame.** When the second batch of plaster has set, invert the frame and remove all its fasteners, including the screw that holds the spacer to the box. Set aside the parts of the frame and gently separate the two halves of the mold. If the parts stick, insert small wooden wedges between the halves to pry them apart; tap the wedges gently with a mallet. When the parts are separated, remove the master (*inset*).

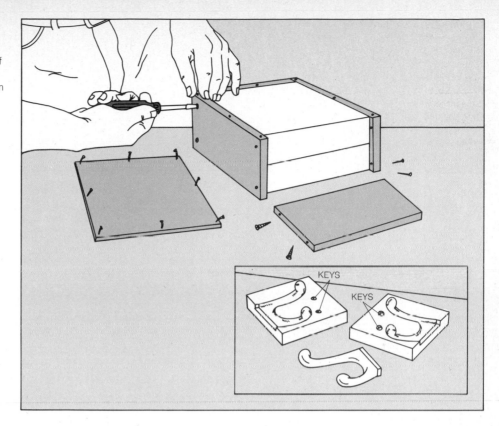

A Three-Piece Mold for Copying a Complex Shape

1 **Creating a clay dam to hold the plaster.** Ring the top of the master with a shallow, saucer-like dam by pressing a ½-inch-wide strip of potter's clay against the horizontal parting line, aligning the top of the strip with the parting line. Cut the strip to size before you apply it, using clay rolled about ⅛ inch thick. If the dam fails to stay in place, buttress it with vertical strips of clay braced against the work surface.

Pinch two ridges in the top of the dam, on opposite sides of the master, to serve as keys in assembling the finished mold. Then coat the top of the master, above the dam, with petroleum jelly, using the technique described on page 70.

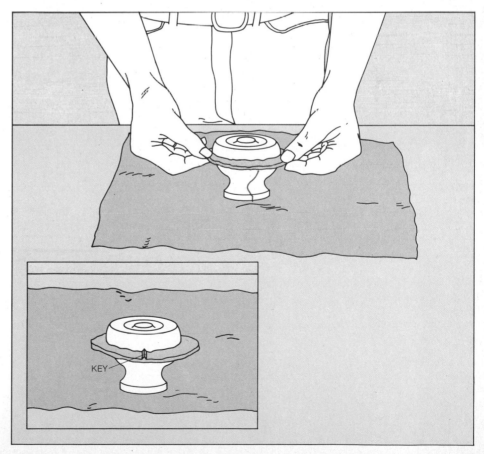

2 Applying the plaster base coat. Fill a wide-nozzled squeeze bottle with plaster mixed to pouring consistency, as described on page 70. Squirt a steady stream of plaster over the top of the master, holding the bottle about 6 inches above the surface and covering the entire area circumscribed by the dam with a thin layer of plaster. When this layer has set, in about 15 minutes, repeat the procedure to build up a double layer of plaster about ⅛ inch thick.

3 Applying the top coat. Mix a batch of plaster and allow it to thicken to molding consistency, as described on page 70. Using your finger tips as a scoop, daub the plaster over the base coat until the combined thickness of the two coats equals about ½ inch, just slightly less than the width of the dam.

When the top coat has set, in about 15 minutes, peel off the clay dam and coat the flat edge of the mold with petroleum jelly.

4 Forming vertical dams. Make two strips of clay ½ inch wide and ⅛ inch thick, as in Step 1, page 73. Press them against the master, aligning them with opposite sides of the vertical parting lines; the master will be divided in two sections, one slightly larger. Pinch a ridge at the midpoint of each strip, for keys in assembling the mold. Put a coat of petroleum jelly on one section of the master between the two dams.

Apply two coats of plaster to the dammed area as in Steps 2 and 3. When the plaster has set, peel off the two clay dams.

5 **Completing the mold.** Spread petroleum jelly on the flat edges of the second mold piece and on the remaining exposed surface of the master. Build up layers of plaster over this surface, as described in Steps 2 and 3, until the third piece of mold is as thick as the other two pieces.

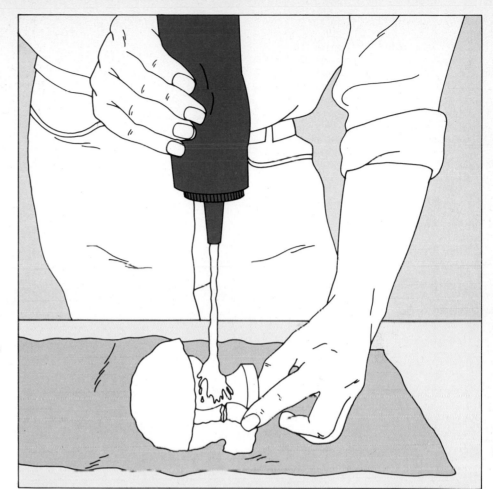

6 **Removing the mold.** When the plaster of the third mold piece has set, separate the mold pieces by driving small wooden wedges into the seams with a rubber mallet. Insert the wedges into the horizontal seam first, placing them at regular intervals around the top mold piece. Gently tap the wedges in succession, driving them to the same depth to equalize their pressure. When the top piece is free, repeat the procedure to separate the sides, alternately driving wedges into both side seams. When the side pieces separate, remove the master (inset).

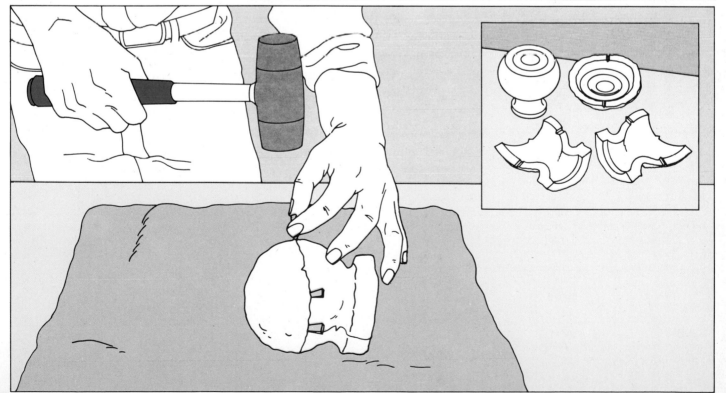

A Wooden Mold for a Fiberglass Shower Base

1 Assembling the sides. To construct the sides of a mold for a shower base, cut four pieces of smooth 1-by-6 lumber, making two pieces as long as the sides of the planned base and two pieces 1½ inches shorter. Miter the ends and bevel the edges of the pieces so that the sides slant about 15° inward from bottom to top. Screw the pieces together to make a four-sided frame, countersinking the screws so that they are below the surface of the wood.

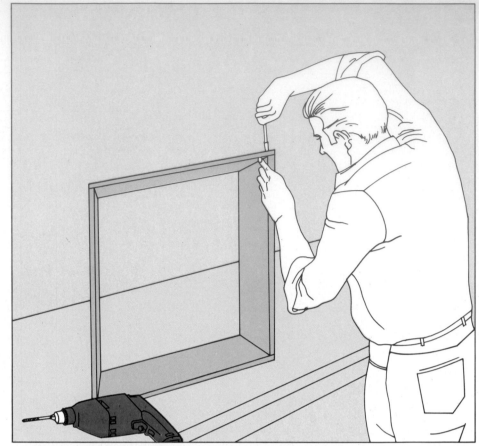

2 Attaching the top. Nail four wood cleats inside the small end of the frame, positioning them flush with the edge of the frame. Cut a piece of smooth ¾-inch plywood to match the interior dimensions of the small end of the frame. Invert the frame, center it over the plywood and, using the cleats as nailing blocks, drive nails through the cleats into the plywood.

Turn the box over and cut strips of quarter-round molding to fit into the angles between the sides and the plywood top *(inset)*. Nail the molding to the sides with finishing nails, then recess the nails with a nail set.

Use a putty knife and wood putty to cover the nailheads and screwheads, fill the seams, and smooth any imperfections in the mold's surface. Sand the entire surface smooth, starting with a medium-grade sandpaper and finishing with a fine grade. Round the corners of the frame, and sand the edges of the molding flush with the top and sides, creating a smooth curved surface over the entire outside of the mold.

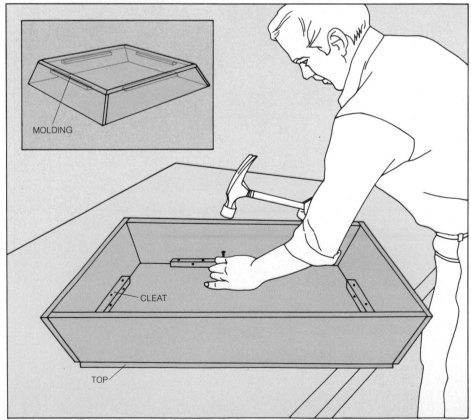

MOLDING

CLEAT

TOP

3 **Attaching the base.** Make a base for the mold with a layer of ½-inch plywood topped with ¼-inch hardboard, 4 inches larger in each dimension than the open end of the frame. Rest the mold on its top, center the base smooth side down on the mold, then screw the base to the frame.

4 **Smoothing the inside angles.** Use potter's clay to round out the interior angles of the mold. Fashion the clay into ropes about ½ inch in diameter by rolling it between the flats of your hands and a sheet of polyethylene. Press the clay ropes into the angles between the sides and the base of the mold, simultaneously shaping them to a smooth arc with the bowl of a spoon.

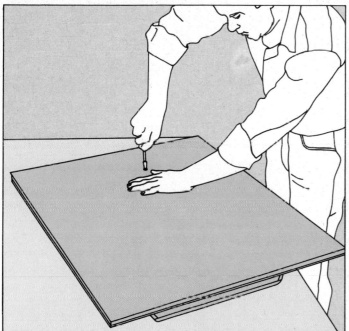

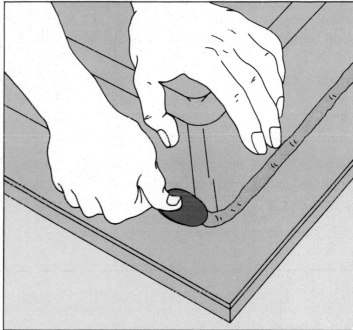

Making Casts of Original Sculpture

Molds can be used for casting original designs as well as for duplicating existing objects. Once you have mastered the fundamentals of mold making, it is a short step to the creation of new shapes for any use you desire.

Plastic castings are particularly well suited for outdoor use, since the finished products resist the elements. You can fashion fanciful adornments—a fountain, a weather vane, a set of outdoor lanterns, or a jardiniere—or replace architectural ornaments such as a finial on a Victorian gable.

Whatever project you choose, the first step is to sculpt or model a master identical in size and shape to the finished product. The simplest, most popular medium for making a master is potter's clay. Small objects can be formed in solid clay; larger objects, and those with projecting parts, require a reinforcing substructure, called an armature.

The armature is built of wood, galvanized pipe or concrete-reinforcing bar roughly in the shape of the object, then covered with wire mesh for the clay to adhere to. For refining and smoothing the master, use the cutting tools, shapers and scrapers sold by art-supply stores for use with potter's clay.

Plan your mold to suit your master as you would for any object (page 70)—with one exception. If you will be making only one casting from an elaborate master and do not care about preserving the master, you can use plaster to make a disposable mold, called a waste mold, instead of using a more expensive flexible mold (pages 78-83).

In construction, a waste mold is similar to a hollow plaster mold; it is made of plaster layered over a master. But the master is always made of potter's clay, and instead of being separated by clay dams, the sections of the mold are

walled off from each other by rows of metal shims pressed into the master along the parting planes. The shims are crimped at intervals to form keys in the edge of the plaster (page 80, Step 2).

The clay master of a waste mold is considered expendable; pieces of it may stick to the mold during dismantling—especially in undercut areas. These pieces of clay are flushed from the mold surface with a stream of water.

Also expendable are the mold sections themselves. When the casting is complete, the plaster is broken away with a chisel and a mallet. To avoid accidentally damaging the casting, work gently and gradually, with shallow cuts. Or do as professionals do: When you construct the mold, add vegetable dye to the first layer of plaster. Then, when you chip off the mold, a change in the color of the plaster will warn you that the casting is not far below.

Flexible Molds to Hold Detail

When you want to cast an object whose contours contain intricate details or an irregular shape, a flexible mold is the best way to ensure faithful reproduction of the original. Similar in principle to a hollow plaster mold *(pages 71-73)* but made with different materials, a flexible mold will bend when removed from the master, so all the eccentric forms and fine details will be preserved intact.

A simple flexible mold for one-time use can be made from plasticine, an oil-base modeling compound *(page 83)*. Molds that are reusable and suited to more complex shapes are made of synthetic rubber, poured in liquid form around the master—the object being replicated—and allowed to cure, or harden. Of the three available types of synthetic rubber—silicone, polysulfide and polyurethane—the least expensive is polyurethane, which can be purchased from plastics suppliers and stores that sell sculptor's materials.

Polyurethane molding compound is prepared by combining two components; the resulting chemical reaction makes the mixture set in six hours. Another 10 hours at room temperature is required for the compound to cure completely. Careful weighing of the components with an accurate scale is essential: An error of 5 per cent in either component can change the consistency of the mold. If the mold is too soft, it may tear; if too hard, it may crack when pulled from the casting.

The pouring consistency of properly mixed molding compound resembles that of thick syrup. It remains pourable for only 20 minutes; therefore, it should be mixed just before you are ready to pour your mold.

To protect the molding compound from moisture, which can keep it from curing fully and make the mold susceptible to tearing, use metal, glass or plastic utensils and containers for mixing. Wooden utensils hold moisture, and so do disposable paper containers unless they are plastic-lined. To sculpt an original master instead of making a copy, use plasticine. When the mold is finished, coat it with petroleum jelly, wrap it in a waterproof covering and store it in a cool, dark, dry place.

The four common types of flexible molds that are shown here and on the following pages allow you to cast almost any shape. Three are of synthetic rubber; one is of plasticine.

The simplest of the rubber molds, ideal for casting flat reliefs, is the one-piece surface mold *(opposite)*. The two-piece shell mold *(pages 80-81)* is the best choice for fragile masters with relatively long vertical axes and varying horizontal diameters. Its external plaster shell lends support to the mold during casting and also reduces cost, since it occupies mold space that would otherwise have to be filled with the more expensive molding compound. The box mold *(pages 82-83)* is excellent for oddly shaped masters, but in making this mold the master is frequently handled; it is not recommended for delicate, easily broken masters.

The plasticine pressed mold *(page 83)*, handy for duplicating small, flat reliefs, can be used immediately, without curing. But the mold yields only one casting because, in being removed from the casting, it must be stretched out of shape.

A One-Piece Surface Mold

1 **Preparing the master.** Rest the master—in this case, a ceiling medallion—face up on a wooden block, leaving at least a 1-inch border, then secure the master to this wooden base with a smooth seal of plasticine. Fashion an open box around the master by building up four walls of plasticine at the edge of the wooden base. Make the walls about ½ inch thick and at least 1 inch higher than the highest point of the master. Coat the master, the base and all four interior plasticine walls with petroleum jelly, applying it as described on page 70.

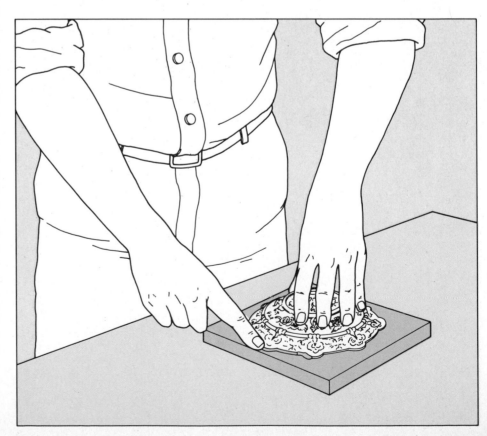

2 **Pouring the mold.** Mix the molding compound, and pour it into the box until it covers the master by ½ inch at its highest point. Allow the compound to cure for 16 hours at room temperature; then remove the plasticine walls, invert the mold and lift off the wooden base.

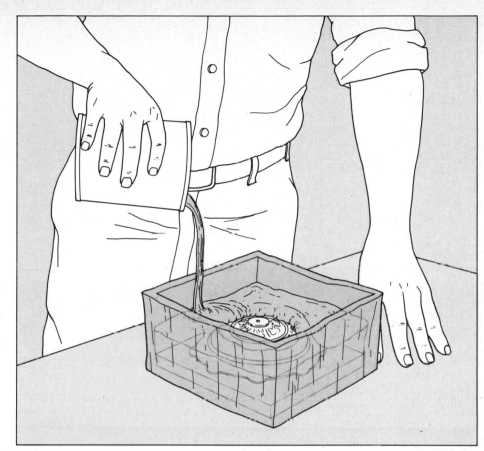

3 **Removing the master.** To loosen the master, grasp the mold on opposite sides and gently bend and twist it (*below, left*). Then place the mold on a flat surface and remove the master by wedging your fingers between the master and the rim of the mold (*below, right*).

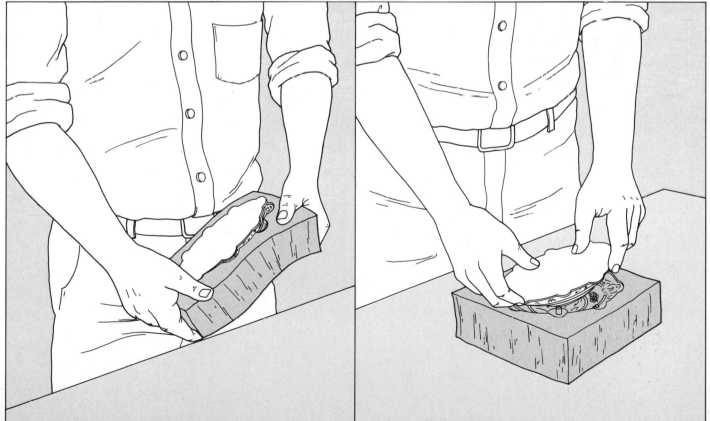

A Shell Mold in Two Pieces

1 Covering the master with plasticine. Seal the master—here, a towel-bar bracket—to a wooden base (page 78, Step 1), leaving a border at least 2 inches wide, then cover the entire surface of the master with a blanket of plasticine ¾ inch thick. Construct a solid column of plasticine at the top of this blanket, about 1 inch high and 1 inch thick, to shape the sprue hole through which the molding compound will be poured.

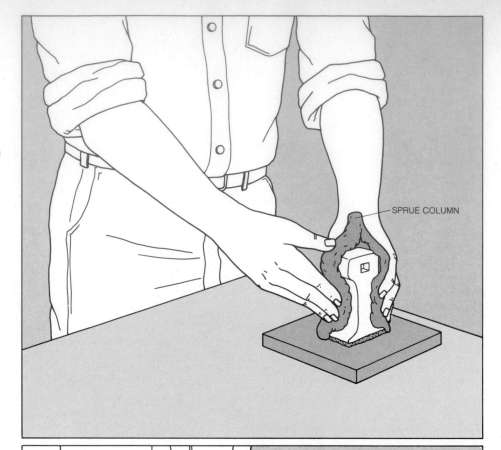

SPRUE COLUMN

2 Implanting shims. Inscribe a vertical parting line down the center of both sides of the plasticine blanket, and press metal shims into the plasticine along the two lines. Cut the shims from disposable aluminum cake or pie tins (aluminum foil is not stiff enough), making them about 2 inches long and 1 inch wide. Angle a pair of shims on each side to make a key in the plaster shell that will eventually encase them (inset). Overlap the shims so that there is no space between them, but do not carry the line of shims over the top of the sprue column.

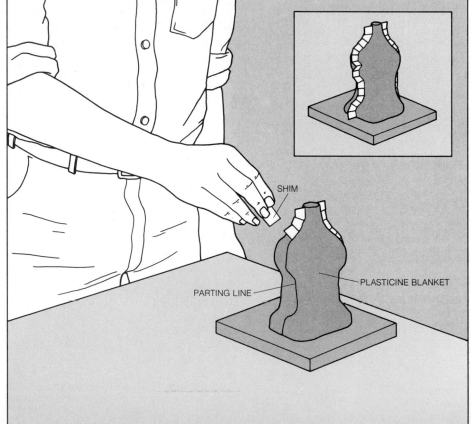

SHIM

PARTING LINE

PLASTICINE BLANKET

3 **Building the plaster shell.** Cover the plasticine blanket with an inch-deep layer of molding plaster *(page 70)*, bringing it level with the top of the shims and the top of the sprue column. Allow the plaster to dry—this usually takes between three and five days.

Mark the outline of the plaster shell on the wooden base, then separate the two halves of the shell along the shim lines. Remove the plasticine blanket and the shims. Coat the master and the inside of the plaster shell with petroleum jelly *(page 70)*. Then reassemble the shell halves around the master, using the outline on the wooden base as a guide. Fasten the shell halves together with reinforced packaging tape, and seal the shell to the base with a bead of plasticine.

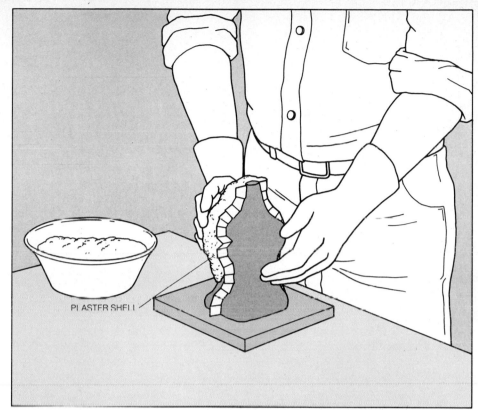

PLASTER SHELL

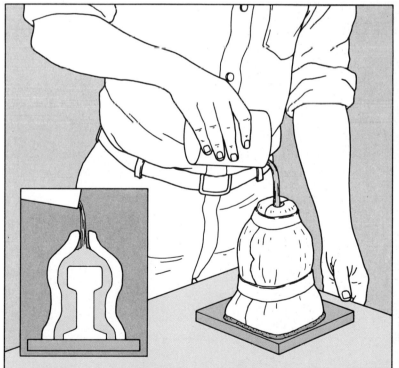

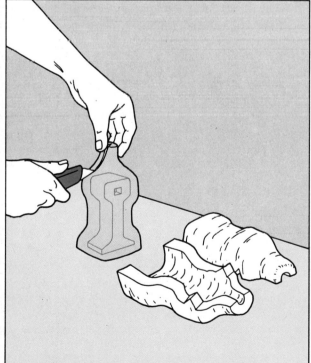

4 **Pouring the mold.** Mix a batch of molding compound, and pour it into the plaster shell until it rises to the lip of the sprue hole *(inset)*. Allow 16 hours for the mold to cure at room temperature, then separate the halves of the shell and remove it. Lift the mold away from the base.

5 **Removing the master.** Use a utility knife to cut through the finished mold, slicing down one side and halfway across the bottom. Gently peel the mold back and remove the master. When casting with the mold, seal the seam with plasticine to prevent leakage.

A Box Mold for Odd Shapes

1 Building a plasticine bed. Cut a wooden base large enough to leave a 2-inch border around the master—in this case, an ornamental door-knob. Place a layer of plasticine on the base and rest the master in it so that the parting line, where the two parts of the mold will separate, runs parallel to the base. Build up a bed of plasticine around the master to the level of the parting line. Then erect a ½-inch-thick plasti-cine wall around the bed and base, carrying the wall about 1 inch higher than the highest point on the master. Press the bed against the wall.

Fashion a roll of plasticine about ½ inch in diame-ter, and wedge it between the master and one wall. Press the roll into the bed to form a half-round berm, defining half the circular sprue hole—through which the liquid resin will be poured *(inset)*. Then circle the master with a V groove, carved into the plasticine bed, to serve as a key for joining the mold parts. Apply a coat of petroleum jelly to the exposed master, the plas-ticine bed and the inside of the plasticine walls, as described on page 70.

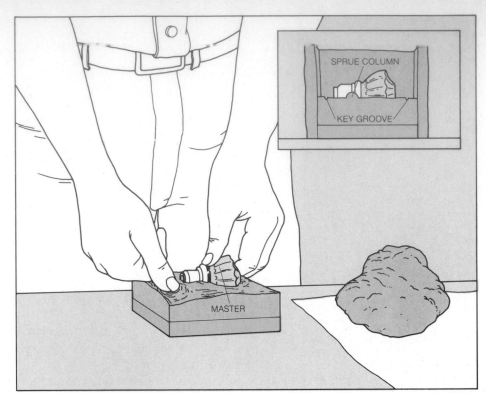

2 Pouring the first piece. Mix a batch of mold-ing compound, and pour it slowly over the master and the plasticine bed until the level is ½ inch above the highest point of the master. When the molding compound has set, in about six hours, invert the entire assembly in preparation for pour-ing the second half of the mold.

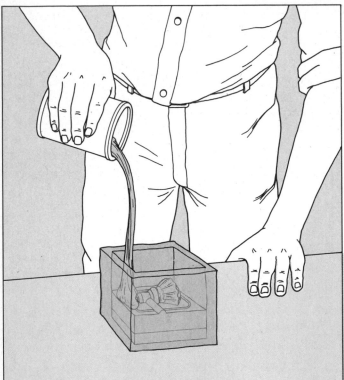

3 Removing the plasticine bed. Lift out the wood base and the plasticine bed, along with the plasticine berm shaping the sprue hole. Repair the wall if necessary. Fit a plasticine roll, ½ inch across, into the half-round sprue-hole depres-sion, to complete the sprue. Coat the master, the mold half and the walls with petroleum jelly.

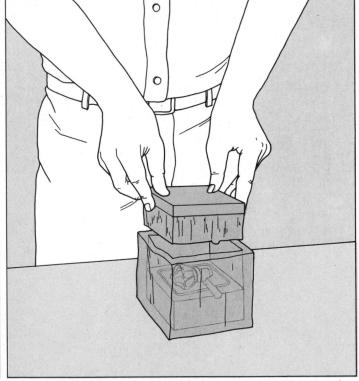

4 **Pouring the second piece.** Mix another batch of molding compound, and pour the second half of the mold to a level ½ inch above the highest point of the master. Allow the compound to cure for 16 hours; then remove the plasticine walls and separate the two halves of the mold. Lift out the roll of plasticine forming the sprue.

5 **Removing the master.** Twist the finished mold to loosen the master. Then remove it from the mold. If this mold will not stand on its own during casting, brace it with two squares of wood, cut to size and taped to the sides of the mold.

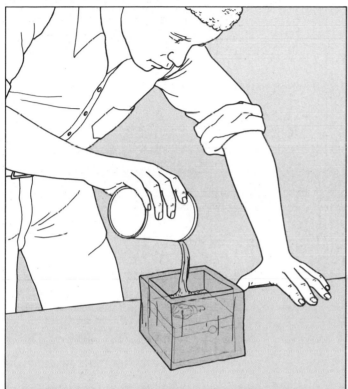

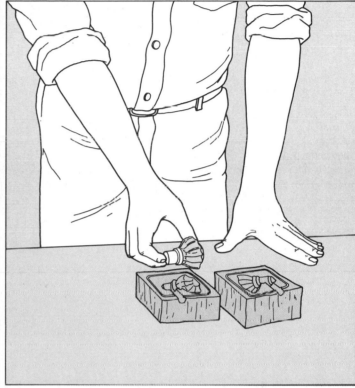

A Simple Pressed Mold for Replacing a Part

Making the mold. Coat the master—in this case, a decorative corner molding—with petroleum jelly *(page 70)* or a light-grade oil, then push an inch-thick slab of plasticine onto it, pressing the plasticine firmly into the detail of the master. Remove the plasticine carefully and check the impression to be sure that all details have been picked up. Lay the mold flat, and cast a duplicate of the master immediately.

Casting in a Mold: Using Resin in Liquid Form

Although a casting acquires shape and texture from the contours of its mold, its weight, color and strength are determined by the casting method and the materials used. Plastic objects can be cast from three liquid resins—epoxy, polyester or acrylic. Each is combined with a hardener that stiffens, or cures, during casting. But the resins differ in cost; in ease of use and in adaptability to the job.

Acrylics have the greatest optical clarity but cost substantially more than other resins. Epoxy resins normally cure to a translucent amber, although they can also be purchased in clearer formulas. They are best used when you want the casting to look like a filler material such as stone dust, powdered wood or powdered metal. Because of their greater adhesive properties, epoxies can hold more filler than other resins can; they can also be combined with a diluent, or thinner, that increases this capacity. Several formulations of epoxy resins and hardeners are available; to be sure of getting the best combination, discuss your needs with the plastics supplier.

Polyester resins are the least expensive, most versatile liquid plastics. Almost as clear as the acrylics, they allow the greatest margin of error in mixing. However, pure polyester resins shrink more than other casting resins as they cure.

When size is critical, you can compensate for shrinkage in either of two ways. For solid castings, you can introduce more resin into the mold by extending the sprue with a collar of clay; as the resin in the mold shrinks, more resin seeps in from the reservoir in the sprue. With hollow castings, in which the resin is painted onto mold sections before assembly, shrinkage along the seams is controlled by the use of a resin containing polyester putty, a dimensionally stable filler that clings to the mold edges.

The casting process begins with mold preparation. A plaster mold must be sealed with shellac to make it smooth and nonabsorbent. Flexible molds need no finishing but, like plaster molds, they must be coated with a mold release or a parting agent, so that the cured casting will slip easily from the mold. The most effective parting agents are polyvinyl alcohol, silicone spray or grease.

To stabilize an irregularly shaped mold during the curing process, add wads of clay or place the mold in sand. Once the cured casting is removed from the mold, it can be finished like any plastic. You can cut off any unwanted projections with a dovetail saw or a hacksaw, and file and sand any rough spots left on the casting, then polish as desired.

During the casting process, work in a well-ventilated area and always use a charcoal-cartridge respirator. Wear rubber gloves and an old shirt with long sleeves. If you get any resin on your skin, use acetone to remove it, then wash with soap and water and apply a skin lotion.

Everything you will need for casting plastic resins is available through plastics suppliers.

Techniques for Mixing and Pouring Liquid Resins

The appearance and performance of a casting made with liquid plastic resins depend on careful measuring and mixing. Mix the ingredients thoroughly, but take care not to introduce moisture or air bubbles.

Ingredients for epoxy and acrylic resins should be measured by weight with an accurate scale. Mix hardener for polyester resins by the drop. The proportion of resin to hardener depends on the characteristics you want. In general, the higher the proportion of hardener to resin, the shorter the curing time; but humidity, temperature and the thickness of the casting also affect how fast or slow the resin cures. To adjust the proportions, mix several test batches under the actual conditions of use, then pick the one that produces the best combination of curing time and physical properties. Keep in mind that the faster the curing, the more brittle the casting—but a casting that cures too slowly may remain rubbery.

To avoid air bubbles in the finished resin, handle the liquid ingredients with care. When preparing to mix them, pour them slowly down the side of the mixing container or mold. Mix gently but thoroughly; if bubbles form, paddle them to the surface. If you are using the mixing attachment of a power drill to combine the ingredients, start or stop the drill with the attachment completely immersed in the resin and run the drill slowly to avoid bubbles.

Make sure that all mixing containers are clean and dry and do not have a wax lining, which could contaminate the resin with wax or moisture. For stirring, use a narrow metal kitchen spatula or a clean, flat wooden stick.

The following guidelines give the correct sequence for mixing most resins, hardeners, pigments and fillers:

☐ RESINS AND HARDENERS. Weigh the resin and hardener separately. Always be sure to pour the hardener into the resin. For polyester resin, add the hardener to the resin one drop at a time.

☐ RESIN AND PUTTY. Pour polyester resin into the polyester putty, then add the cream hardener supplied with the polyester putty. Measure the hardener by squeezing specified lengths from the tube onto your stirrer.

☐ ADDING PIGMENTS. To color acrylic and epoxy resins, add pigments a little at a time to both resin and hardener until each is the desired color, then pour the hardener into the resin. To color a polyester resin, add the pigment to the resin only, then add hardener. To color a polyester resin-and-putty mixture, first combine the putty and resin, then add the pigment and finally the cream hardener.

☐ ADDING FILLER MATERIAL. Add powdered wood, stone dust, or powdered metal to polyester resins and putty resins following the same sequences used for adding pigments. With epoxies, thin the resin first with a diluent, add hardener, then add filler.

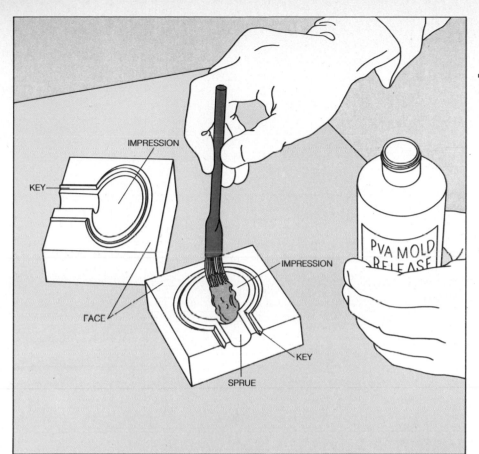

Pouring a Solid Casting in a Flexible Mold

1 Applying the mold release. Using a small nylon- or natural-bristled brush, cover the entire inside of both halves of the synthetic-rubber mold—in this case, for an oval doorknob—with a coating of polyvinyl-alcohol mold release, including the face and key as well as the mold impression and sprue. When the alcohol in the mold release has dried, leaving only a slick surface, apply a second coating. If there are undercut areas of the mold that are difficult to reach with a brush, pour a small amount of the mold release directly into the mold and swish it around; then pour out the excess. When the second coat of polyvinyl alcohol has dried, join the two parts of the mold by fitting the keys together.

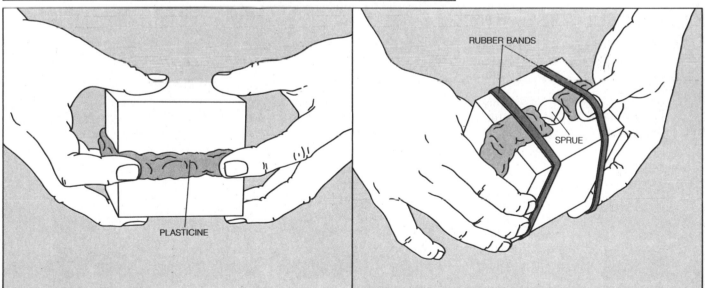

2 Sealing the mold. With the two halves of the mold held securely together, press a strand of plasticine modeling clay, about ¾ inch in diameter, along the seam line (*above, left*), sealing off all but the sprue hole through which the resin will be poured. For convenience, work with small lengths of clay, overlapping their ends by about 1 inch. Then snap wide, flat rubber bands around the two halves of the mold at intervals of about 1 inch (*above, right*). Do not let the rub-

berbands cover the sprue hole, and do not make them so tight that they warp the mold or press it out of alignment. Snap a second set of rubber bands over the mold, perpendicular to the first set and at the same intervals.

Mix resin, hardener, pigments and fillers according to the manufacturer's instructions and the guidelines opposite. Be sure, at this point, to put on a charcoal-cartridge respirator.

3 **Filling the mold with resin mixture.** Hold the mold slightly tilted, sprue hole up, and pour the resin mixture slowly into the sprue. Let the resin slide gently down into the mold, following the contours of the casting impression; this will reduce the number of air bubbles that form.

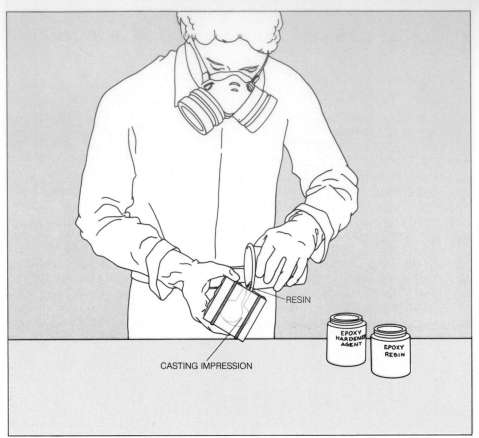

RESIN

CASTING IMPRESSION

EPOXY HARDENER AGENT

EPOXY RESIN

4 **Removing air bubbles.** Hold the filled mold at an angle, sprue up, and tap it lightly several times against the work surface. Turn the mold one quarter revolution and, with the sprue still angled up, tap it against the surface again. Continue turning and tapping the mold until you return to the original position. Set the mold aside, sprue up, and allow the casting to cure.

Use the resin left in your mixing cup to determine when the casting has cured. When the surface of the resin in the cup is no longer tacky to the touch, press your finger firmly against it. The casting has cured when your finger no longer leaves an impression. Remove the casting from the mold by taking the rubber bands and plasticine off and pulling the two halves of the rubber mold apart. One half will still be clutching the casting. Simply peel the mold from the casting. Remove any defects and finish the casting as described on page 89.

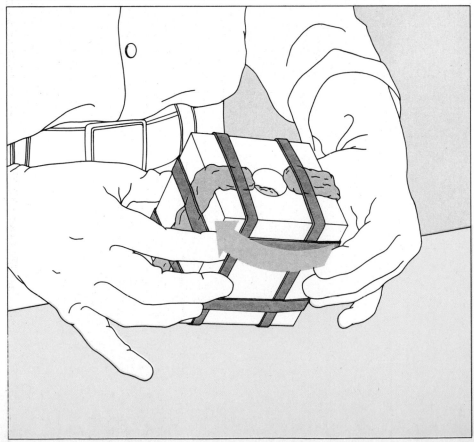

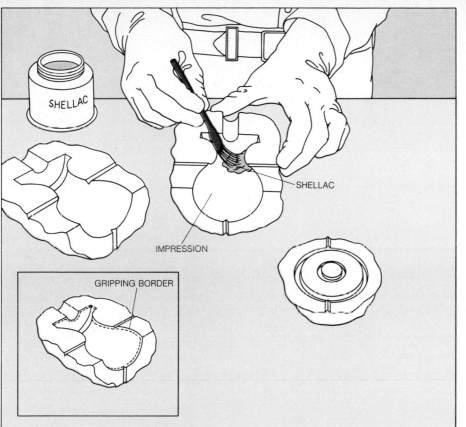

Making a Hollow Casting in a Rigid Mold

1 **Preparing the mold interior.** To cast the hollow knob used in this example, coat the inside face of each part of the plaster mold with shellac, using a small brush with nylon or natural bristles. Handle the mold pieces gently to prevent chipping. When the shellac is dry, in six to eight hours, spread mold release over the impression, as in Step 1, page 85, but stop ⅛ inch short of the edge of the impression (*inset*). The resulting border allows the resin to grip the edges of the mold pieces while curing, and thus reduces shrinkage along the seams.

Mix a casting compound consisting of three parts polyester putty to one part polyester resin; use a cream hardener and add any pigments desired. Be sure to wear a charcoal-cartridge respirator during the mixing process.

SHELLAC

IMPRESSION

GRIPPING BORDER

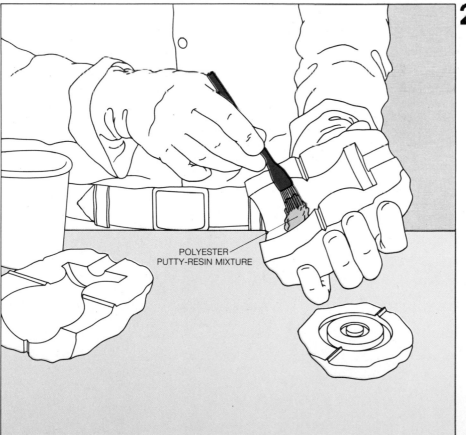

2 **Coating the mold with resin.** Use a small brush to spread the polyester putty-resin mixture evenly into the impression on each piece of the mold. Cover the impressions completely, including the gripping border, with a coating about 1/16 inch thick. When this coat has partially cured to a gelatinous consistency (usually after about 10 minutes), mix another batch of polyester putty and resin, then brush on another coat. Mix and add a third layer when the second has partially cured, building up a combined thickness of about ¼ inch.

At this point, if your casting is large and demands reinforcing, lay resin-soaked pieces of fiberglass mat into the impressions of the mold. Be careful to keep the mat pieces well within the borders of the impressions.

POLYESTER PUTTY-RESIN MIXTURE

3 **Assembling the resin-coated mold.** Fit the parts of the mold together in proper alignment and secure them temporarily with rubber bands or light-gauge wire. Then drill holes sideways through the seams of the mold with a high-speed twist drill bit, spacing the holes about 4 inches apart. Drive wood screws—use a size slightly larger in diameter than the drill bit—through the holes and then take out the temporary fasteners.

If the mold seams are not thick enough to support screw holes, fasten the parts of the mold together with light-gauge wire, twisting the ends of the wire to tighten it against the mold *(inset)*.

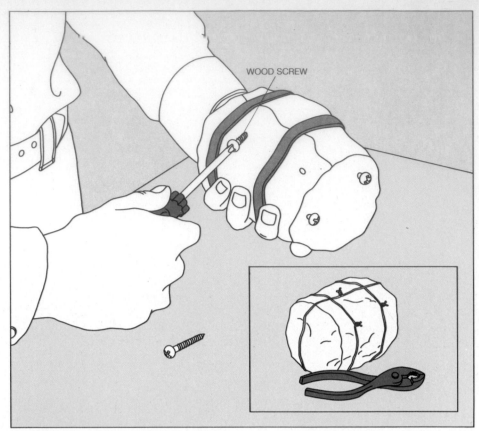

WOOD SCREW

4 **Adding a bonding coat of resin.** Mix a second batch of resin and pour it into the mold through the sprue hole. Swirl it around so that it coats the inside of the casting and fills any hairline cracks at the seam lines; pour out the excess. When the first bonding coat has hardened, add further coats if desired, increasing the thickness of the casting by about $\frac{1}{16}$ inch each time.

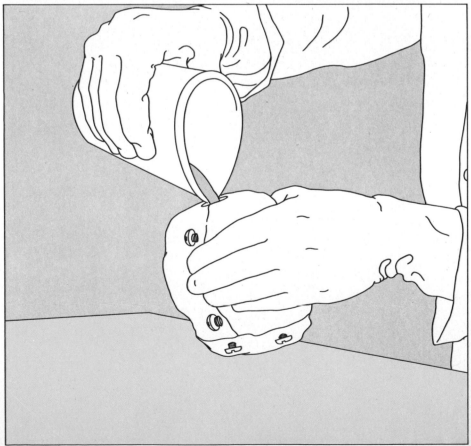

5 **Reinforcing the seams.** While one of the internal bonding coats is still gelatinous, push a piece of resin-impregnated fiberglass mat through the sprue hole *(below, left)*. Spread the fiberglass and press it against the inside of the casting with a wooden dowel *(below, right)*. If the casting you are working with is a large one, insert additional pieces of fiberglass for extra reinforcement. Lap the fiberglass over the seams of the separate pieces of the mold, thoroughly bonding them together.

When the casting has hardened, remove the screws or the wires holding the mold together. Gently separate the pieces of the mold, pulling them away from the casting. If the mold does not separate easily, hammer lightly on wooden wedges to gently pry it apart along the seams.

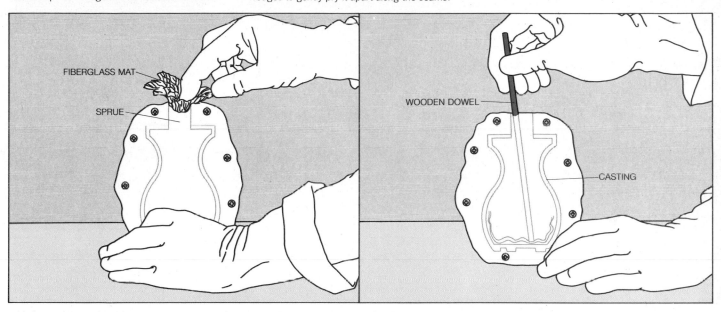

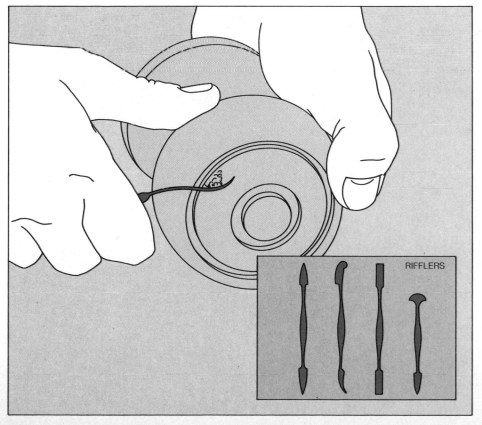

Refining the Details of the Finished Casting

Removing surface imperfections. Use the rasplike faces and edges of small woodworking tools called rifflers *(inset)* to carve or shave away blemishes acquired during the curing process or while the casting was being removed from the mold. If a defect is extensive, restore it with a resin patch *(page 107)*, and use rifflers to sculpt detail into the patch. A small chisel or utility knife can also be used for this purpose.

Glass Fibers Reinforced with Plastic Resin

Few building materials can match the versatility and utility of home-mixed glass-fiber-reinforced plastic, commonly called fiberglass. Strong, lightweight and completely weatherproof, it can be molded into free-form shapes, such as basins or furniture, or laminated to make decorative wall panels, sills or roofing. Formed of woven or felted glass fabric embedded in a plastic resin, the material combines the best attributes of both. The strength of glass fibers in the fabric reinforces the weak, brittle plastic; the plastic makes the material rigid and gives it a smooth, impermeable surface.

For an even stronger material, the proportion of glass to resin is increased within a range of 30 to 70 per cent of the total weight. The glass content is determined by the density and arrangement of the glass filaments in the fabric and by the amount of resin used to laminate it.

The fiberglass fabrics used in laminating (below) are all made up of rovings, bunched strands of glass filaments. The rovings are pressed or woven into fabrics of varying densities; density is measured in terms of the weight, in ounces, of 1 square foot of fabric. The fabric weight, as well as the forming method, determines the quantity of resin needed. Generally, a chopped-strand mat takes double its weight in resin; the ratio of resin to fabric for a woven scrim is one to one.

Fabric choice depends on the amount of reinforcement needed and on the type of resin to be used. The bonding agents in fiberglass fabrics are designed to bond with different resins; in purchasing the fabric, be sure to specify the resin you have selected. Because woven fabrics, which are stronger, do not bond as well with each other as they do with mat fabrics, it is best to alternate layers to ensure even strength throughout the laminate. And for a smooth surface, a fine fiberglass mat is often used for the layer just beneath the final coat.

Although many resins are suitable for making fiberglass, polyester is both the simplest to use and the most economical, and it provides excellent strength and moisture resistance. When laminated, polyester's normal shrinkage of 10 to 15 per cent during curing is reduced to 2 per cent or less, and this slight shrinkage is often an advantage: In molded laminations, it allows the finished object to slip easily from the mold. Polyester resin is usually available as a two-component system, with resin in one container and hardener in another. Be sure to follow the manufacturer's mixing instructions exactly, combining the components carefully to avoid mixing in air bubbles, which weaken the cured resin. For detailed information on measuring and mixing resins, see page 84.

Usually the resin includes an accelerator, which speeds the curing time, but sometimes this component must be added separately. If you are adding an accelerator, be absolutely sure to mix it into the resin before introducing the hardener. Mixing the accelerator with the hardener can result in an explosive chemical reaction between the cobalt naphthanate in the accelerator and the peroxide in the hardener.

A special resin called the gel coat is always used as the outermost layer in fiberglass lamination. This unreinforced resin provides a smooth, glossy, protective layer between the glass fiber and outside moisture. It is applied as the first layer if the fiberglass is being built up in a mold, but as the final layer on a flat lamination. By using a tinted gel coat, either premixed or mixed on the job with up to 10 per cent of a suitable polyester tinting paste, you can impart a surface color to the fiberglass.

Gel coats come in several formulations. An air-curing type is easiest to handle; it is viscous and less likely to run or, when dry, to crack. Ideally, it should be applied

A Glossary of Glass Fabrics for Laminating

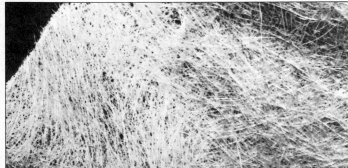

Chopped-strand mat. The most common fiberglass reinforcement is a feltlike fabric of short, randomly arranged glass strands held together by a binding agent. Standard 1.5-ounce mat is well suited for making reinforced fiberglass roofing sheets and wall panels. Mat fiberglass is stiff and difficult to form until the binding agent dissolves in the resin.

Woven roving. Excellent for reinforcing walls, joints and corners, woven roving provides great strength. Because the coarse weave tends to cause air bubbles to form in the resin between layers, woven roving is usually alternated with layers of mat to prevent delamination. The coarse weave also makes woven roving unsuitable as a surface layer, since the weave shows through the surface resin. Woven roving in weights greater than 2.18 ounce is too stiff to be used with ease in home workshops.

with a brush in a single coat. If it is the first layer, it should be thoroughly cured before the laminating resin is applied; otherwise, the solvent in the laminating resin could react with the gel coat and damage its finish. The gel coat should also be compatible with the laminating resin; information about compatibility is included on the label on the can.

When you are laminating fiberglass with a mold or form from which it will later be removed, you will need a release agent to keep the fiberglass from sticking. For simple molds, you can coat the mold surface with lacquer spray or cover it with a sheet of polyethylene film. For large or complex molds, a two-stage parting compound of hard wax and polyvinyl alcohol is generally used; it is available from plastics suppliers. Apply the wax first and let it dry completely; then apply the alcohol, using a sponge to lay on a thin, even coat. Protect the surface from dust while the parting compound dries.

Cut all the fiberglass fabric to the proper size before mixing the resin. Most such fabrics can be cut with scissors or a utility knife, or they can be torn by pulling them over the edge of a handsaw. If the cut fabric edges are to be lapped, comb or fray the edges of the glass strands to flatten them so that they will intermingle within the resin and make a joint without a visible seam.

To impregnate the fiberglass with resin, use a paint roller or a brush, stippling the brush over the fabric so as not to dislocate the glass strands with the sticky, resin-coated bristles. Never apply more than 6 ounces of resin over a square foot of fabric; the heat generated by the resin as it cures can adversely affect the laminate if it is applied in quantities too great.

Some polyester resins cure completely only in the absence of air; left exposed, they remain tacky indefinitely. This can be an advantage in multilayer lamination over a large area, where new coats are usually applied before the previous coat cures—a forming process called wet-on-wet. The work can be interrupted for several hours without harming the laminate. When using such resins, however, the final coat must be sealed with a gel coat or covered with an airtight layer of polyethylene film until the resin cures, after which the film can be peeled off. Or the final coat can be painted with acetone, which dissolves the uncured resin but leaves a rough surface.

Because working with resin is messy, use disposable utensils for measuring and mixing it whenever possible. Keep tools soaking in solvent so that you will be able to clean them when the job is completed, before the resin hardens. A supply of clean rags is essential for removing spilled resin, and the floor beneath the work should be covered with newspapers to catch drips, which are difficult to remove once they have had time to harden.

In addition, the laminating process also requires several specialized tools. One is a metal disk roller for consolidating the resin and glass and pushing out air bubbles that form between layers. Rollers come in several forms, with washer- or paddle-shaped blades, and in sizes ranging from ¼ inch to 12 inches wide. For contoured surfaces there is a flexible roller, with a head resembling a length of coiled spring. To trim and finish the completed lamination, you will need a metal-cutting saw, a forming tool, and a supply of wet-or-dry sandpaper graduating from 240- to 600-grit.

Most of the components used in laminating fiberglass are irritants. Always wear a respirator and goggles when cutting or sanding fiberglass to protect yourself from fine glass fibers released into the air, and protect your skin with gloves when you handle the material. Both the resin and its hardener are caustic, flammable and toxic. Work with them in a well-ventilated place, away from flame, and keep them away from skin and eyes.

Because of the large quantities of resin often involved, laminating is best done outdoors, where the fumes are less dangerous. But do not work outdoors when temperatures are below 65° F., or the resin will cure too slowly. Also, avoid direct sunlight and hot or cold drafts, which might adversely affect the curing process.

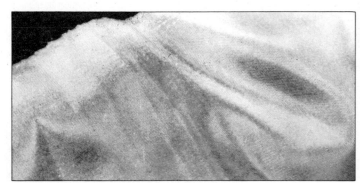

Glass scrim twill. Finer and more flexible than standard woven roving, scrim twill is a good choice for covering irregular shapes and compound curves. Standard .55-ounce scrim twill offers excellent resistance to stress, and is often alternated with mat for high-strength laminates.

Surface mat. A feltlike material, finer than ordinary mat, surface mat ranges in weight from .6 ounce to 1.1 ounce. It is used as a smooth final layer to conceal the texture of coarse fabrics.

Weatherproofing a Table for Use Outdoors

1 Cutting fabric to fit. Cut .55-ounce scrim twill to cover the top and edge of the table, adding 2 inches along each edge to lap to the underside. If the tabletop is too large to cover with one piece, cut two pieces with a ½-inch overlap; then fray the overlap until the strands of fabric can be intermingled and pressed flat. Tack the fabric to the table edge near each corner, then cut away a square of fabric at each corner so that the fabric will fold flat at the corners; again allow for about a ½-inch overlap around each corner, and fray the overlap so that the strands can be pulled around the corner.

Remove the tacks and carefully lift the fabric from the table. Using an orbital or belt sander with coarse abrasive paper, sand the tabletop and edges to remove all paint, dirt and grease; round all the edges. Countersink any screwheads or boltheads, and fill all holes with wood putty.

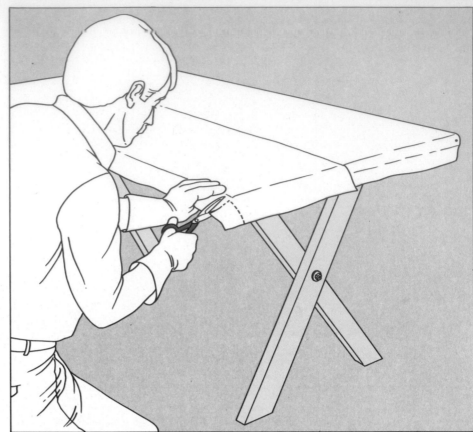

2 Priming the surface with resin. Using a paint roller, apply a thin primer coat of resin to the top and edges of the table and to 2 inches of the underside. Mix the resin according to the manufacturer's instructions, then thin it to the consistency of paint by adding a bit of acetone.

Allow the primer coat to harden partially (about 45 minutes) while you clean tools and mixing containers with acetone. When the primer coat starts to harden, mix another batch of resin but do not thin it. Then, using the roller, apply a second coat over the primer coat.

3 **Laying the fiberglass fabric.** Have a helper hold one end of the precut fiberglass fabric away from the resin-coated table while you align the other end with one corner of the table. Then lower the fabric into the resin, lapping it over the edges and underside and intermingling the frayed strands at the corners. Avoid wrinkles; if the fabric must be moved, lift it and realign it. When the fabric is in place, stipple it with a resin-filled soft-bristled brush until it is completely flat. For a large table, repeat this process, covering the other half of the table with fiberglass. Overlap the frayed edges of the two pieces of fabric by ½ inch, flattening the frayed strands so that the seam line will be invisible.

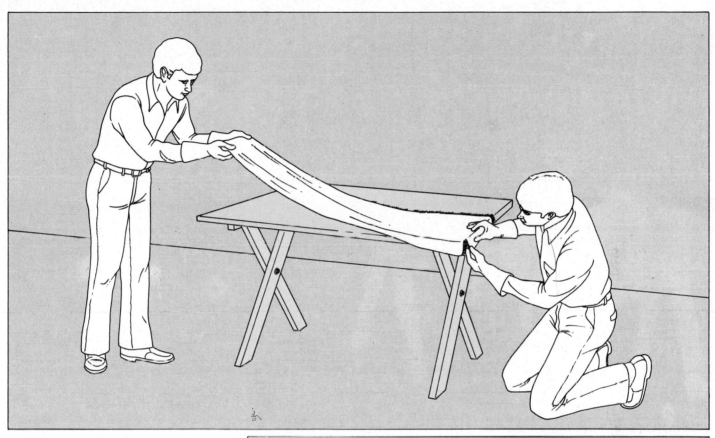

4 **Rolling down the fabric.** Force the fabric into the resin with a disk roller *(inset)*, working from the center of the table out to the edges. Roll out any air bubbles, then check to be sure that all the fabric is saturated with resin. Let the resin cure for two hours; meanwhile, clean the tools.

Cover the laminate with a presealing coat of resin, adding tinting paste *(page 90)* if you want color in the final surface. Use a soft-bristled paintbrush to lay on the resin in an even coat, about ⅟₁₆ inch thick. When this coat has hardened completely, sand down any uneven spots with 320-grit emery paper.

Mix enough air-drying gel coat to cover the laminate in one thin application. If you require more color, you can add up to 5 per cent tinting paste. Apply the gel coat in the same way as you applied the presealing coat.

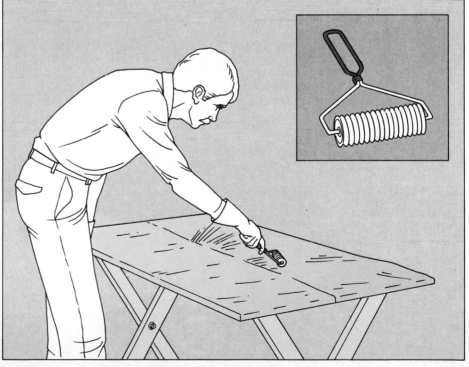

Molding a Base to Use in a Shower Stall

1 **Making a pattern.** Cut from brown wrapping paper a pattern that will be large enough to cover the mold. Then shape this paper pattern to the mold, cutting away the excess paper at the corners but allowing for an overlap of ½ inch wherever possible.

Use the paper pattern to cut four identically shaped pieces of fiberglass fabric: two of 1.5-ounce mat, one of .55-ounce twill, and one of thin surface mat. Cut the fabric with sharp scissors to the exact outline of the pattern. Fray the overlapping edges as in Step 1, page 92.

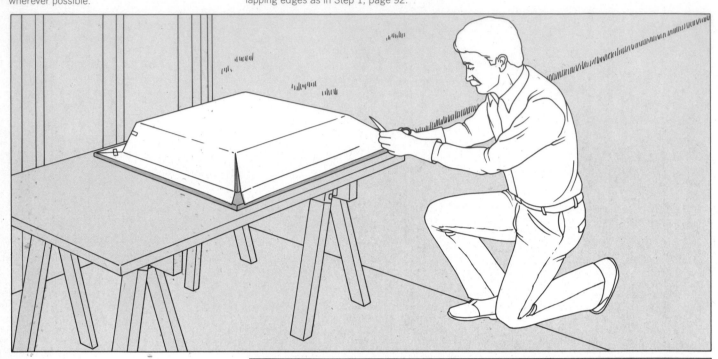

2 **Applying the parting agent.** Use a rag to apply a coat of paste wax on the surface of the laminating mold; polish the wax by rubbing in a circular motion with a lint-free cloth. Apply and polish two more coats of wax. Then use a square-edged plastic sponge to apply a thin, even coat of polyvinyl alcohol over the entire mold. Work with smooth strokes, always in the same direction, and overlap successive strokes to avoid gaps and streaking.

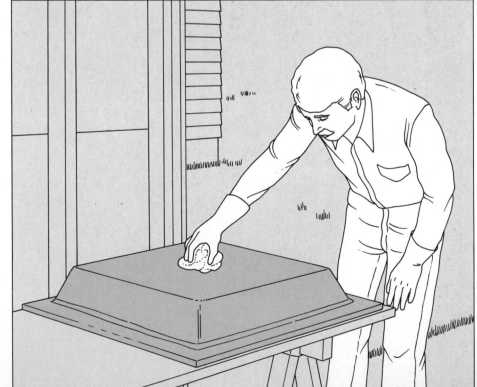

3 **Laying on the laminate.** Mix a batch of air-drying gel coat sufficient to cover the entire mold. Use a soft-bristled paintbrush to apply the gel coat to the mold and to a small piece of scrap wood. When the gel coat on the scrap wood is tacky (usually in about 30 minutes), mix a batch of laminating resin, enough that it will cover the entire mold.

Using a paint roller with a ½-inch nap, coat the mold with laminating resin. Then lay the pre-cut fiberglass surface mat over the mold, using a helper to align it *(page 93, Step 3)*. Work the fabric into the resin with a soft-bristled brush and remove air bubbles with a disk roller *(page 93)*, taking special care to flatten the loose strands at the corners. Before the resin cures, apply successive coats of resin and fabric in the same manner, starting with a layer of 1.5-ounce mat, then a layer of twill, and finally a second layer of the 1.5-ounce mat. Compress each layer and roll out air bubbles before adding the next coat of resin.

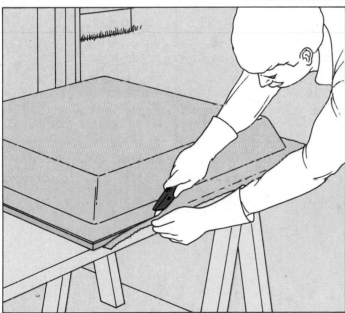

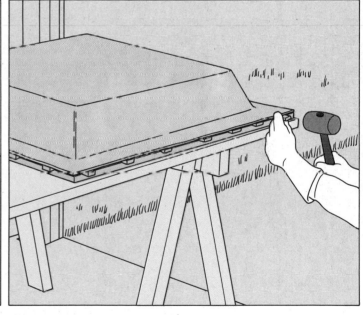

4 **Trimming the finished form.** When the final coat of resin has become rubbery (after about 45 to 60 minutes), trim away the excess laminate along the mold edge with a utility knife; do not lift or bend the laminate. Let the laminate cure overnight or longer, following the manufacturer's instructions, in a warm, dry, dust-free place.

5 **Releasing the mold.** Insert thin wedges of wood or plastic between the laminate and the mold at 6-inch intervals. Tap the wedges gently with a mallet. If the laminate does not spring free, do not try to pry it loose. Instead, invert the mold and run warm water into it to melt the wax parting compound.

Finish the rough edges of the laminated basin by filing them with a forming tool, then sanding with 600-grit emery paper. Working from the inside of the basin, drill a drain hole with a carbide-tipped hole saw fitted in a power drill. Put a wood block under the drain-hole area while drilling, to avoid splintering the laminate.

Tracking Down Fiberglass Flaws

The many variables involved in fiberglass lamination can cause some problems in working with the resin and some flaws in the finished product. In most cases you can trace the source of the problem fairly easily and correct the damage.

Errors in measuring and mixing the resin can produce a resin mixture that does not harden. This may be the result of adding too much or too little hardener, or of using resin or hardener that is too old and no longer effective. If the resin fails to set on only part of the work, probably it has been unevenly mixed, or the mold or parting compound was not allowed to dry completely. If resin does not cure after 3 to 5 hours, you will have to scrape it off, thoroughly clean the surface, and mix new resin.

A workshop that is too cool can also delay curing: Keep the temperature at 65° or above. But a room temperature that is too high—above 85°—can cause the resin to cure too quickly or to harden in the mixing pot before it is applied. Mixing too large a batch of resin can have the same result. Activated resin generates heat as it cures, and large quantities produce enough heat to make a significant difference in hardening time. Left open too long—around 45 minutes—a large quantity of activated resin may even begin to smoke, becoming a fire hazard.

Some errors are apparent only after the laminate has cured; they manifest themselves as shown in the photographs at right. Most can be remedied by the application of a second gel coat over the surface. However, in the case of leaching, it is necessary to cut away the flawed area and replace it with a fiberglass patch (page 111). The same is true of any area where the layers of fiberglass fabric separate—or delaminate—and are no longer solidly bound together with resin.

Common Surface Flaws and How They Are Caused

Crazing. Fine hairline cracks may appear in the surface immediately or as much as several months later. They are caused by incorrect proportions in the resin mixture or by the use of a laminating resin that is not compatible with the gel coat being used (page 91).

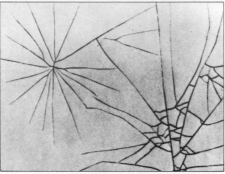

Star cracking. Radiating cracks may become visible in the gel coat if it has been applied in too thick a layer or if the underside of the laminate receives a sharp blow.

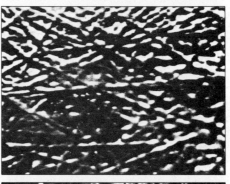

Fiber pattern. A raised honeycomb effect on the laminate surface is the result of failure to use a layer of fine surface-mat fiberglass between the gel coat and thick woven roving. If the surface is smooth but the fabric pattern is visible, the gel coat is too thin or, in a molded form—where the gel coat is applied first—the fabric was laid on before the gel coat was stiff enough.

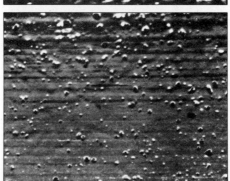

Pinholing. A pocked surface results from the introduction of tiny air bubbles into the gel coat, usually during mixing. Bubbles are particularly difficult to see in a tinted gel coat. Also, if the gel coat is too viscous, the bubbles, though visible, are difficult to remove.

Leaching. Fiberglass fabric is left exposed when resin is washed away, either because it was improperly cured or because the resin used is not weatherproof. The fragility of the exposed glass in a leached area substantially weakens the laminate; the laminate should be patched.

Molecules Custom-built to Serve a Specific Need

Most of the materials used for construction are nature's gifts and can be manipulated only within fixed limits. Woods can be seasoned and formed, metals tempered and alloyed, stone crushed and shaped, but the basic qualities of each material are inbuilt and unchangeable. Because they are synthetic, however, plastics can be tailor-made: Not only the end product but the very character of the substance can be shaped to its purpose.

Leo Baekeland, a Belgian-born chemist working in New York, was the first scientist to create a truly made-to-order material—earlier versions of plastics, such as celluloid (page 7), had consisted of chemically treated natural substances. In 1907, searching for a substitute for shellac, Baekeland mixed and heated two unpromising-looking liquids, phenol and formaldehyde, both smelly derivatives of coal. When his flask cooled, he found a resin altogether unlike its ingredients and far superior to shellac. He had produced the first synthetic plastic, which was later trade-named Bakelite in its creator's honor.

Today, such metamorphoses of familiar substances or their extracts into synthetics possessing durability and ease of shaping not to be found in nature have become commonplace. Most of them involve a few colorless gases and a volatile liquid or two derived from petroleum or natural gas. Using a combination of heat, pressure, and catalysts—substances that hasten chemical change without taking part in it—industry converts these raw materials into the plastics that ease modern life at every turn.

The secret of these remarkable transformations, from Baekeland's successful simulation of shellac to the routine triumphs of today, lies in the invisible realm of molecular structure. A typical plastic molecule strings together hundreds or thousands of carbon atoms, each surrounded by a complement of other atoms. This chain of atoms, although at best only a few billionths of an inch long, is a molecular giant.

Despite their size and complexity, these giant molecules are relatively easy to assemble. Petroleum and natural gas provide molecules containing only a few atoms, mostly carbon and hydrogen, often arranged like very short sections of the immense carbon chains of plastics. The magic of technology can cause these short sections of chain to line up and fuse together end to end, each of them adding several links to the larger chain. In the language of science, each small molecule is called a monomer; the result of their linking, the unwieldy but invaluable plastic molecule, is a polymer.

In its simplest versions, the plastic polymer consists of a single kind of monomer, repeated hundreds or thousands of times over. Polyethylene, the waxy, pliant stuff of squeeze bottles, is a scrambled mass of such polymers. The monomers from which they are made are, as the name of the plastic indicates, molecules of ethylene gas.

Scientists can shape the properties of a plastic not only by choosing the monomers, but also by controlling the way these building blocks are arranged at the crucial moment when the monomers unite chemically to form the plastic polymers. In basic, or so-called low-density polyethylene, for instance, numerous side branches trail off the main carbon chains. These branching chains act as spacers, separating the molecules and reducing their tendency to cling to one another. This polyethylene is flexible and lightweight; it softens at moderate heat.

For a polyethylene that withstands higher temperatures, chemists have developed a way to assemble unbranched polymer chains. With no side chains to space them, these molecules pack more closely, cling together more tightly and make up a denser, less flexible plastic with greater heat resistance. Tougher still is a version with links between adjacent polymer chains. Created by drenching the plastic in high-energy radiation, these cross links bind the molecules in a rigid matrix, giving the familiar squeeze-bottle polyethylene enough durability for aerospace uses.

The invisible intricacies of plastics. Each of the six-atom monomers in the polyethylene molecule (below, left) has two atoms of carbon (in color) and four of hydrogen. Before they are strung together, these monomers make up explosive ethylene gas. But when thousands of them fuse, they form a molecule of resilient plastic. Many-branched chains (below, center) yield a pliant polyethylene that softens at 200° F. Linked in a latticework (below, right), the same chains constitute a polyethylene that can withstand twice that temperature.

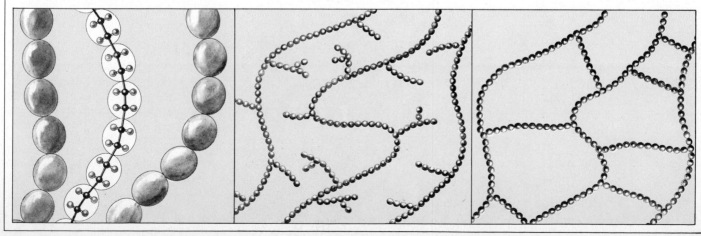

Laminated Surfaces Move out of the Kitchen

The same plastic laminates that work so well in the kitchen and bath as easy-care coverings for countertops are equally useful in other rooms of the house, where their sleek surfaces can cover coffee tables, window sills and bookshelves. Mindful of this decorative potential, manufacturers are making the laminates in bold colors, ranging from bright scarlet to deep navy blue; in wood-grain patterns that faithfully reproduce the graining and color of everything from light oak to Burmese teak; and even in simulations of burlap, slate and leather.

Laminates can be successfully applied over existing laminated surfaces and over virtually any smooth wood. One of their chief attractions is that they permit you to build pieces from inexpensive hardboard, particle board or plywood. Laminates cannot, however, be applied over linoleum or ceramic tile without an underlayment of wood, although it often is possible to avoid this intermediate step by using laminated panels—large sheets of plastic laminate that come already bonded to hardboard.

Regardless of the surface to which they are applied, plastic laminates are usually fastened in place with one of two adhesives. Both technically are contact cements, which means that the cement is spread on two surfaces, and bonds immediately when the surfaces are brought together. One of the cements is water-based and easy to use; it is also nontoxic and nonflammable. But its bonding strength is low, so it is best used for large, flat surfaces. The other cement, far stronger, is petroleum-based and, like all such products, is hazardous. It is toxic, extremely volatile and explosive.

Before opening a container of this petroleum-base cement, extinguish all smoking materials and pilot lights, and flip the circuit breakers for fans, refrigerators and other motor-driven equipment in the work area. Do not let the cement touch your skin, and avoid breathing its fumes. Work in a well-ventilated area, preferably outdoors—although outdoor temperatures can interfere with drying time. The cement bonds best at about 70° F. Hot weather may make it dry too fast, and in cold weather it may dry too slowly and become brittle. Either condition weakens the bond.

Both forms of contact cement are applied in the same way, with a paintbrush or a roller. Normally one coat is enough, but some wood surfaces—especially the end grain of plywood—will soak up the cement and require a second coat. When the cement has dried to the tacky point, the two surfaces are ready to be joined.

Use care in positioning the laminate; once the two layers of cement have touched, it is virtually impossible to pull them apart. Edging strips and similar small pieces can usually be guided into place without any preliminaries, but larger pieces may need prior alignment. Since the tacky cement sticks only to itself, you can keep the two surfaces separate during alignment by covering them with strips of wood or pieces of wax paper until they are properly positioned. As soon as the laminate is aligned, slip out the strips of wood or the wax paper and press the laminate into the adhesive by rolling it with a roller of wood or hard rubber, or by passing a wood block over the laminate while repeatedly tapping the block with a mallet.

Most laminates come in 4-by-8-foot sheets, though larger sheets—up to 5 feet wide and 12 feet long—are obtainable on special order. The standard sheet is $\frac{1}{16}$ inch thick, but sheets $\frac{1}{32}$ inch thick are sometimes used for surfaces that get little wear, such as the backs of cabinet doors.

A special laminate, called postforming grade, is useful for installations where the laminate is required to fit over a sharply curved surface. The paper layers in this laminate are not the customary kraft paper; instead, they resemble crepe paper. For wide curves, such as those on the edge of a round or an oval table, postforming laminate can be bent while it is cold. For tight curves, it must be warmed. In factories, where industrial heaters and presses are available, interior curves with a radius as small as ¼ inch are possible; but when the laminate is heated at home, curves with a radius of less than ¾ inch are impractical. To warm the laminate, a heat gun or a heating iron, tools normally used to remove paint, is ideal.

When applying a laminate to any surface, old or new, make sure the surface is scrupulously clean, smooth and dry. Sand down any rough spots and fill in any indentations with wood putty. If you are relaminating over an old laminate, cement any loose edges. Break any surface bubbles in the old laminate by hitting them with a mallet. Remove hardware, sinks and faucets from kitchen and bathroom countertops, and sand the old laminate to roughen its surface so that the cement will adhere better. Be sure, however, to clear away any chips or sanding dust. These not only interfere with the bond but may telegraph their presence in the form of lumps in the finished surface.

The laminate itself is frequently cut as much as ¼ inch larger than the surface it covers and then trimmed to size after it is in place. The preliminary cutting techniques for plastics are described on pages 20-29. For a neater finish between two laminated surfaces that meet at right angles, the edge is usually beveled. Both the trimming and beveling operations can be done with an electric router or with hand tools—specifically, a flat cabinet rasp and a No. 8 mill file.

Seating a vertical edge. Brush an even coat of contact cement on the laminate and on the vertical surface being covered. When the adhesive is tacky, grip the edges of the laminate in both hands, nestling the edges in the last joints of your fingers so that your finger tips are free to use as a guide in aligning the top edge of the laminate with the top of the vertical surface (inset). Then bring the two adhesive-coated surfaces together, pressing the laminate against the vertical edge, and roll the laminate with a small wooden or rubber roller until it is firmly seated.

Rounding a sharp curve. Apply an even coat of contact cement to the laminate and the surface being laminated; join them as far as the point where the curve begins, using the same techniques shown at left. Then don gloves and warm the laminate with a heat gun or heating iron, holding the heat source just above the plastic. When the plastic is pliable—usually in about 30 seconds—pass it quickly around the curve and press it into the adhesive.

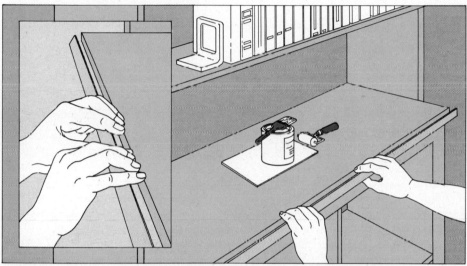

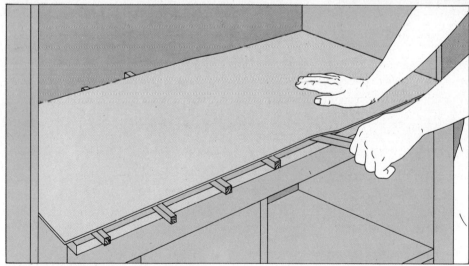

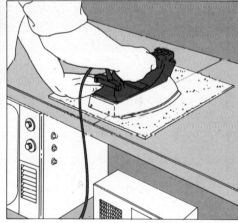

Aligning a large surface. Apply contact cement to the laminate sheet and to the horizontal surface being laminated, using a little extra cement around the edges. When the adhesive is tacky, lay wood strips at about 1-foot intervals across the horizontal surface; the strips should be at least ¾ inch thick. Place the laminate in position on top of the strips. Starting at one end, remove the wood strips one at a time, pressing down the laminate as you go. When the two surfaces are joined, roll the laminate or use a wood block and a mallet to set it firmly onto the horizontal undersurface.

Making a seam. To join two pieces of laminate, trim both edges to meet precisely (pages 21-22). Apply contact cement to the laminate pieces and the surface being covered. Join one laminate piece to the base as shown at left. Then align the second piece, beginning at the seam. Press both pieces into the adhesive, using a roller or a mallet and a wood block to seat them firmly. To strengthen the seam line, cover it with a cloth and run a household iron over the plastic until it is warm to the touch; use the lowest setting on the iron. This will soften the cement, embedding the plastic more securely.

Choose an inconspicuous location for a seam. Do not allow the seam to run into an opening, such as the cutout for a sink. Avoid joining small sections of laminate; such seams have a tendency to pull away from the base.

Finishing the Edges of a Laminated Top

1 Trimming an overhanging edge. If the plastic laminate has been cut slightly larger than its base, remove the overhang with an electric router or a flat cabinet rasp. Fit the router with a flush-trim cutter, and pass the cutter along the overhanging edge (*below, left*), successively shaving off narrow strips of plastic until the cutter's roller guide comes in contact with the vertical surface at a right angle to the top laminate (*in-*

set). Then make one slow final pass, using the vertical surface as a guide, to cut away the last of the excess plastic.

When using a rasp (*below, right*), cut away the excess laminate with light downward strokes. Hold the rasp perpendicular to the edge of the laminate, but move it at a slight angle, covering 2 or 3 inches of overhang with each stroke.

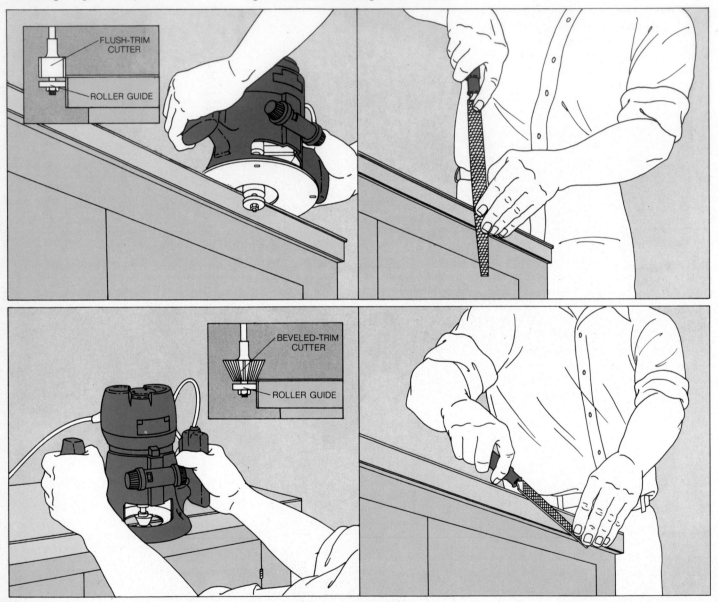

2 Beveling an edge. Put a beveled-trim cutter on the router and adjust the depth of the cut, setting the cutter so that the bottom edge of the blade falls just slightly below the bottom surface of the laminate being beveled (*inset*). Switch on the router motor, and when it reaches full speed, bring the cutter gently in contact with the plastic. Move the router along the edge in

one smooth pass (*above, left*), keeping the roller guide in constant contact with the underlying vertical surface.

To bevel the edge by hand, use a No. 8 mill file, holding the file at about a 25° angle to the edge and moving it downward in angled strokes that cover 2 to 4 inches at a time (*right*).

Sheathing Three-Dimensional Objects

Laminated objects intended for less utilitarian purposes than kitchen countertops require preliminary planning in order to ensure neat, unobtrusive joints. The beauty of laminates is only skin deep. When their patterned or colored surfaces are cut, the core is exposed, and it will be visible as a dark line wherever two pieces of laminate meet at right angles. But there are ways to minimize these lines—by plotting the sequence in which the object is assembled and its surfaces are covered. The assembly of the parson's table at right shows proper planning.

Before applying a laminate, first consider the angle from which the object will most often be viewed. The surface facing the viewer is the one that should be covered last. On a low table, for instance, the sides should be covered before the top. The dark line of the core will then face the sides of the table, where its presence will be less obvious. On the other hand, the sequence of placement should be reversed for a high shelf such as the mantle over a fireplace, because the top of the mantle tends to be in view less often than the sides.

The second consideration in planning neat joints is the nature of the surface over which the router's roller guide passes as the router trims or bevels laminated edges. The smoother this surface, the neater the beveled or trimmed edge. Wherever possible, try to plan these finishing operations so that the router will move over the side grain of plywood or over a previously laminated surface. Running the router over plywood end grain, with its bumpy surface, will inevitably create a wavy edge.

A Table Intended for Laminating

1 **Cutting the base and top.** Cut a square tabletop from ¾-inch plywood. Cut two side-pieces the same width as the top, and two endpieces 1½ inches narrower. The endpiece legs will thus be ¾ inch narrower than the sidepiece legs, but the aprons of all four pieces will be identical in width. Lap the sidepieces over the endpieces and secure them with six penny (2-inch) finishing nails. Nail the top to the resulting skirt assembly.

Cut eight plywood rectangles for the inside legs. Each rectangle should be the same height as the skirt assembly, but four should be as wide as the outside legs of the endpieces and four should be ¾ inch narrower. Nail the wide and narrow rectangles together in pairs to form each L-shaped interior leg. Then join the L-shaped legs to the skirt assembly, nailing through the face of the skirt into the interior edges of the legs. Countersink all nails.

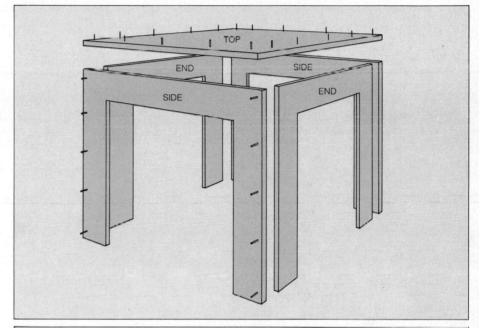

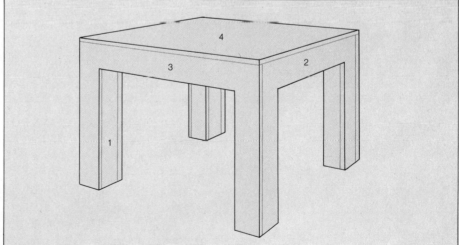

2 **Completing the lamination.** Using the techniques described on pages 98-99, apply laminate in the numbered sequence shown here. First cut eight rectangles of laminate to fit the inside surfaces of the legs. Trim the rectangles even with the bottom of the end and side skirts, and apply them to the insides of the legs (1). Then sheath the endpieces of the skirt assembly (2), thus covering the end grain of the plywood of the sidepieces; trim the laminate to size either with a router or with a rasp and a file, as shown opposite. Next cover the side-pieces (3), trim the laminate to size, and bevel the outside corner of each leg. Finally apply a piece of laminate to the top (4), trim it to size and bevel the top edges.

4

A Wealth of Fillers and Paints

Convoluted cover-up. Colorless epoxy, called liquid glass, is swirled over a butcher-block table-top, protecting its surface while allowing the beauty of the grain to shine through. The pourable plastic, roughly the consistency of molasses, will be smoothed with the edge of a plastic-covered playing card to form a moisture-proof, scratch-proof seal ⅛ inch thick. With successive applications, the coating can be increased to ¼ inch without altering the color or clarity of the material that lies underneath.

Patching, resealing and recoating are the inevitable by-products of having a home: No household surface takes care of itself. Exterior walls, roofs and walkways will weather, crack and change shape under attacks of wind, rain and temperature fluctuations. Interior walls, ceilings, floors and built-in fixtures submit to nicks, scuffs and regular washings, not to mention an occasional blow severe enough to cause greater damage.

In times past, such household repairs often took more time and effort than they were worth. It was easier to live with the scuffs and cracks than to fix them. When restoration was necessary, it was frequently left to professionals, who possessed the specialized tools and equipment—and the years of practice—needed to do the job well. But plastics have changed all that. Today's easy-to-use synthetic fillers, caulking compounds, sealants and paints make most kinds of surface repairs and refinishing a simple and economical matter, even for the amateur. The same is true for the use of these materials on new work.

The qualities that set these plastics compounds apart from traditional materials are their superior powers of adhesion and their infinite shapability. True to their name, they can be formulated to respond to virtually any demands made of them.

Troweled-on polyester metal fillers, for example, will bond to irregularly shaped holes and dents, eliminating the need for costly welding. Silicone and polysulfide caulking compounds will remain permanently pliant, outlasting their oil-base predecessors, which in time become brittle. Quick-setting synthetic fillers and coatings can be applied without special tools and are packaged in small quantities for household-sized jobs. One-coat synthetic-rubber waterproofing emulsion eliminates the need for expensive layer-over-layer built-up roofing. Fast-curing epoxy mortars and grouts are far stronger than conventional cement and plaster, and new plastic paints and varnishes cover any surface you are likely to want to finish or refinish. There is even an aquatic epoxy paint that goes on underwater, for refurbishing swimming pools.

Properly applied, most of these synthetic fillers and paints produce surfaces more durable than the natural materials they cover. But when they are used to repair leaks and cracks, more than resurfacing may be required. Sometimes there are serious structural defects that must first be diagnosed and remedied. The success of the job will also depend on choosing the right coating, one that is compatible with the surface beneath, and on preparing that surface properly. These important preliminaries are discussed in this chapter, along with the techniques for applying the new plastic coatings.

Compounds to Patch Holes in Almost Anything

Plastic filling compounds are used to patch holes in everything from teeth to auto fenders. Strong, fast-setting and resistant to water and chemicals, they are ideal for a wide range of repairs around the home. Whether the job calls for filling a gouge in a kitchen countertop, patching a small hole in a water tank or shaping a replacement for a broken corner of a lawn chair, the right plastic filler and application technique can produce a nearly undetectable repair.

The polyester or epoxy resins that are the base for such plastic filling compounds as wood putty or metal paste can be used in their clear form or mixed with filler materials and pigments to match almost any surface. The choice between epoxy and polyester should be based on the job at hand. Epoxies are more expensive, but they are stronger, adhere better to many surfaces, and offer greater resistance to heat and corrosive chemicals. Unless these properties are essential, a polyester resin is usually sufficient; polyesters match epoxies in water resistance, are more tolerant of mixing errors, and cure more quickly. Problems of polyester adhesion can usually be overcome by the simple process of roughening the surface or drilling anchor holes.

Both polyester and epoxy filling compounds are available ready-mixed or in multiple-component products. A ready-mixed filler, such as the familiar plastic wood putty, can be used directly from the tube or can; the resin, hardener, filler and color are already blended. Most such ready-mixed compounds are sold in small quantities and are practical for touch-up work or small repairs, although they are more expensive per ounce than multiple-component fillers.

The parts of a multiple-component filler must be precisely measured and mixed, following the same procedures used for mixing casting plastics (*page 84*). With some of these products, you add the hardener to factory-mixed resin and filler; with others, you add a factory-mixed paste of hardener and filler to the liquid resin. If the hardener and filler are in paste form, mix the paste first with a small quantity of resin to liquefy it and prevent lumps from forming when the rest of the resin is added.

Some products require you to mix fillers and pigments with the resin before you add the hardener. Because of the increased potential for error with such a product, you should make trial mixtures before you mix the batch you will use on your repair. Pay particular attention to the proportion of filler material to resin. Too little filler will result in excessive shrinkage as the mixture cures; too much filler will make the mixture crumbly and reduce its adhesion.

Filling compounds can be matched in color or other characteristics to a wide variety of surfaces by the addition of appropriate filler material. Metal powders, for example, can be used for cosmetic repairs as well as in applications where resistance to stress or friction is required. An iron filler is the one that offers the best resistance to heat or impact, whereas an aluminum filler is preferable if the repair will be subject to friction. Graphite powder added to a filler improves its sliding properties.

Nonmetallic filler materials are also available, including ceramic powders, chopped glass fiber, and powdered stone. Glass fiber is an all-purpose filler that is also a strong reinforcement, which makes it useful for sheet-metal repairs. Since it is water-resistant when cured, it is well suited for repairing rust holes on water tanks, gutters, steel panels, or even cars. Thinned with acetone to the consistency of paint, glass fiber can also be used to prime bare metal before further repairs are made with chopped glass or other filler.

Fillers of powdered ceramic, which have good insulating properties, are the best to use for repairs where heat or electricity is present. Such ceramic fillers can be readily tinted by adding colored powders, which are available at most paint or hobby stores.

For patching and repairing wood, many ready-mixed compounds are available in a variety of wood colors. These tinted fillers cannot be further colored with stain; however, they can be retinted before they are applied to achieve the exact color desired, or touched up with an oil-base paint after they have cured. You can avoid the necessity for this chore by mixing your own filler, using a polyester res-in base and fine sawdust from the wood you are using. If you intend to stain such a mixture at the same time you are staining the surrounding wood, use a high concentration of sawdust in the filler, and brush the hardened patch with acetone to expose as much of the sawdust as possible prior to staining.

You will need a kitchen scale to measure ingredients by weight, or a glass measuring cup if the ingredients are to be measured by volume. Use disposable cups and spatulas for mixing, or use glass and metal utensils that can be cleaned with acetone to prevent impurities from getting in the filler. Clean paintbrushes with acetone between applications, or use very cheap brushes and discard them after they have been used.

An electric drill is handy when you are preparing a surface for repairs. Fitted with a sanding disk, it can be used to remove paint or corrosion quickly; with a burr bit, it can be used to grind a feathered edge around a hole. Masking tape and paraffin are needed to contain the filler and to protect the surrounding surface from solvents.

To put a smooth surface on the filler, you will need to cover it with a sheet of clear plastic household wrap, which will not stick to the filler. To polish a finished repair, use a buffing disk with a plastic-polishing compound (*page 106*).

Tips on Safe Handling

Because almost all of the chemicals in plastic fillers are volatile and somewhat caustic, you must take care to ensure adequate ventilation and avoid open flames. Use plastic gloves and protective eye goggles to prevent accidental contact with the materials. If any resin does get on your skin, use the recommended solvent to remove it, then wash the affected area immediately with soap and water. Before opening any of the filler containers, read the manufacturer's safety information and familiarize yourself with the remedies prescribed.

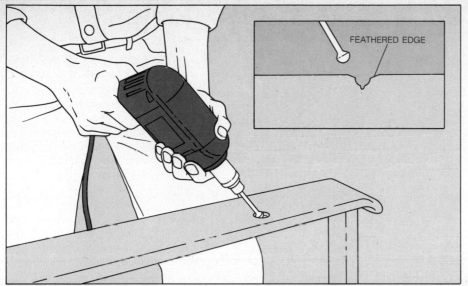

A Nearly Invisible Patch for a Small Hole

1 Preparing the hole. With a burr attachment on an electric drill, feather the edge of the hole at a shallow angle to a distance of ¼ inch *(inset)*. Use the drill's cooling-fan exhaust to blow loose particles and dust from the hole.

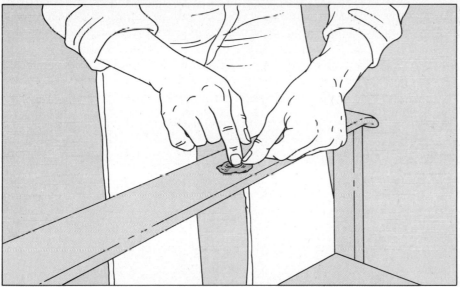

2 Making a wax dam. Shape a bit of paraffin wax with your fingers to build a ⅛-inch dam around the hole, slightly outside the feathered edge. Do not let the wax touch any part of the hole; it would keep the filler from sticking.

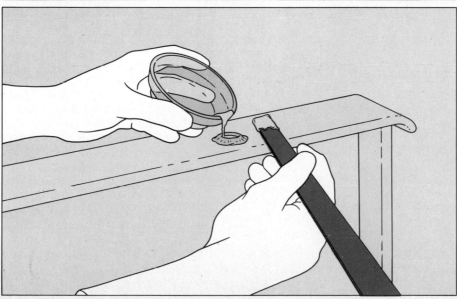

3 Filling the hole. Using a shallow dish, mix enough filler to fill the hole, then use the mixing spatula to drip filler into the hole. Tamp the filler with the spatula to force out any air bubbles. Then add filler until it is about ¹⁄₁₆ inch above the undamaged part of the surface.

4 Trimming off the excess. When the plastic filler has partially hardened (usually in about five minutes), use a sharp, wide bladed putty knife or a single-edged razor blade in a holder to shear away the wax dam and the excess filler flush with the surface. Avoid gouging the filler or the wood beside it with the corner of the blade.

Discard any leftover filler and clean the tools with acetone. Mix a small batch of fresh filler.

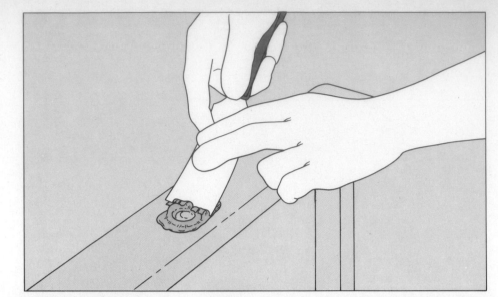

5 Finishing the filler. Drip a small amount of filler onto the repair and cover it with a piece of clear plastic wrap. Spread the filler evenly by lightly smoothing it with a single-edged razor blade drawn across the film. Remove the film and allow the filler to cure for two to five hours, according to the manufacturer's instructions.

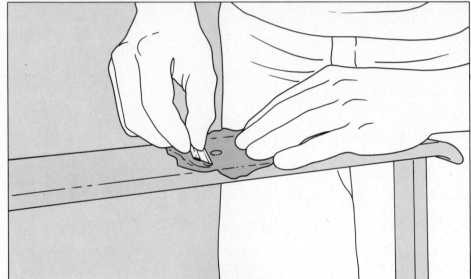

6 Buffing the repair. After lightly hand-sanding the patch with 400-grit emery paper, spread rubbing compound on a buffing attachment for an electric drill and buff the patch, using light pressure and a circular motion. Continue until the patch area blends into the surrounding surface.

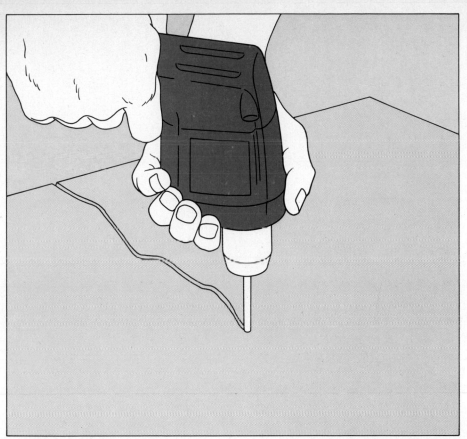

Stopping a Crack and Sealing It with Filler

1 **Drilling to stop the crack.** Drill a hole ¼ inch in diameter through the cracked plastic surface to relieve the tension that is creating the crack. When you do this, use the end of the crack as the center point of the hole.

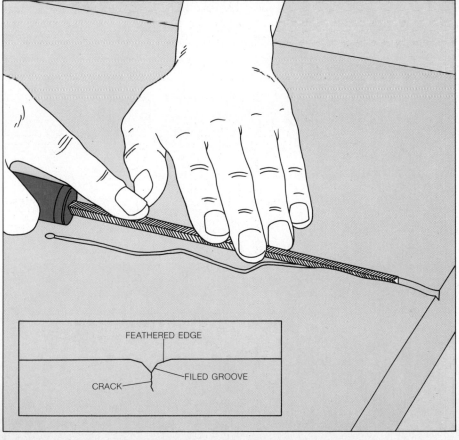

FEATHERED EDGE

CRACK

FILED GROOVE

2 **Widening the crack for filler.** Use a small triangular file to enlarge and deepen the crack, making a V-shaped groove at least ⅛ inch deep. Feather the edges of the grooves slightly (*inset*), and clean out all loose particles and dust. Cover the undamaged surface with masking tape extending right to the feathered edges.

3 **Filling the crack.** Using a shallow dish, mix enough plastic filler to fill up the filed groove, then apply the filler with a putty knife, working across the groove. Build up the filler about $1/16$ inch higher than the original surface. When the filler has partially hardened, remove the masking tape and complete the repair as shown in Steps 4 to 6, page 106.

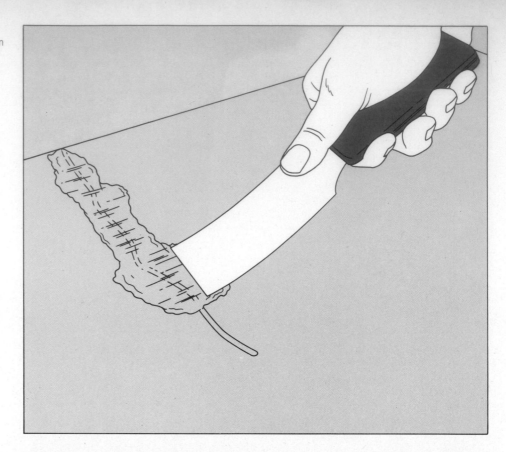

Anchoring Filler in a Sheet-Metal Dent

1 **Stripping paint from the surface.** Remove paint and other protective coatings from the damaged surface, using an electric drill fitted with a wire-brush attachment. Using heavy pressure and a slow, sweeping motion, expose shiny metal across the entire damaged area and 1 inch more of the surrounding surface in all directions. Use either detergent and water or alcohol to clean and degrease the stripped metal.

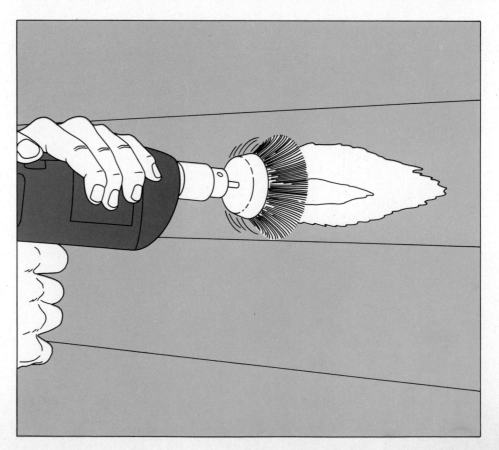

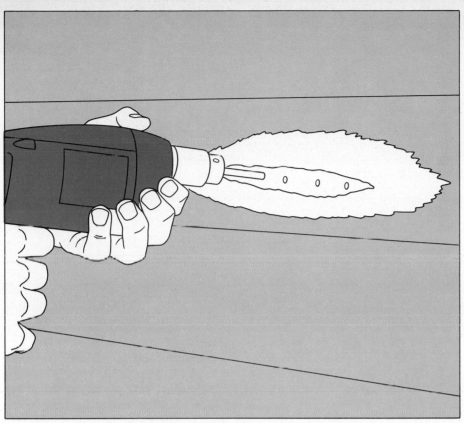

2 **Drilling anchor holes.** Drill ¼-inch anchor holes about 1 inch apart within the damaged area to offer a good grip for the filler. Paint a thin primer coat of polyester resin over all bare metal.

Cover the area around the primed surface with masking tape. When the primer is no longer tacky, fill the dent with polyester filler (*Step 3, opposite*). Let it cure, then sand it smooth and paint it to match the surrounding surface.

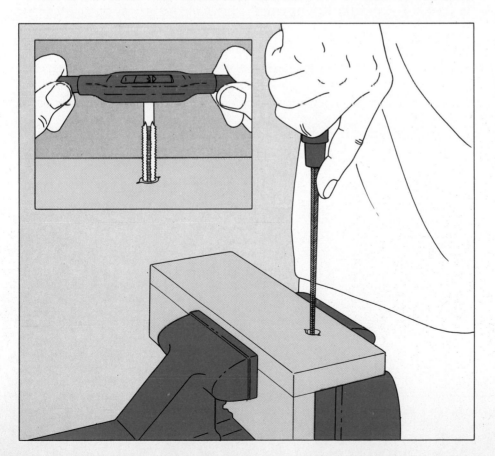

Renewing Stripped Threads with an Anchored Plug

Preparing the hole. Remove the damaged threads by drilling out the hole with a bit one and a half times the original diameter of the hole. Use a small triangular file (*left*) to cut several notches in the side of the hole; these notches, which should be at least ⅛ inch deep, will prevent the filler plug from rotating. Clean the hole and degrease it with alcohol, then fill it as shown in Step 3, opposite, using a filler that matches the color of the surface.

Allow the filler to cure completely (this will generally take from about two to five hours), then redrill the hole through the filler plug to the original diameter. Cut new threads in the new hole with a tapping tool (*inset*).

Watertight Fiberglass Patches

A gaping hole in a plastic, metal or wood surface that once was smooth and watertight need not spell disaster. Using a variation of the fiberglass-lamination techniques shown on pages 90-96, you can apply a strong, watertight patch to anything from a plastic laundry tub to a fiberglass roof panel. Careful sanding, finishing and tinting can make most such repairs virtually undetectable.

Both the polyester and epoxy resins used in working with fiberglass will also bond well with wood and metal. However, polyester's relatively short drying time, low cost and easy workability make it preferable to the more expensive epoxy-base resins; the latter adhere more readily but are harder to use because they are so thick. For most household repairs, the best choice is a preaccelerated, two-component polyester resin with a liquid hardener, available from plastics suppliers and most marine-hardware stores.

Fiberglass repairs are based on the same principles as fiberglass lamination: Overlapping layers of resin-impregnated cloth form a solid surface across the damaged area. Careful mixing of the resin compound ensures proper bonding between the layers of glass and the material being patched. A gel coat applied to the working side of the patch creates a smooth and waterproof surface. (For a discussion of components and mixing techniques, see page 84.)

Most resins are mixed with 3 per cent hardener by weight. Cool weather slows curing; if necessary, you can increase the proportion of hardener 5 per cent to speed curing, but be aware that excessive hardener can cause brittleness by curing the resin too quickly. In some cases you may be able to speed curing by warming the area of the repair with a heat lamp. If repairs are urgent in temperatures below 54° F., you can use a technique known as cross-hardening, which allows curing down to 32°. Cross-hardening involves first mixing the standard hardener into the resin, then adding a second hardener containing benzoyl peroxide. Ask for specific information on proportions if you purchase materials for cross-hardening.

To get a strong patch, you must prepare the surrounding surface carefully. Make sure it is free of loose particles, dirt, paint and grease. If the surface is metal, scrape around the hole with a wire brush until the metal shines, then degrease the metal with soap and water, or with alcohol or vinegar. When the metal is dry, prime it with a thin coat of plastic filling compound (page 108).

Since resin-saturated glass fabric is slippery and awkward to handle, it is convenient to soak the fiberglass in resin, then transfer it to the repair on a sheet of clear plastic household wrap. This keeps the glass fabric from coming apart and minimizes dripping.

Because fiberglass work is messy and stray resin is extremely difficult to remove if given time to cure, use disposable containers and utensils if possible. Otherwise, clean the equipment between steps with acetone or another recommended solvent. Cover the floor and surfaces adjacent to the work area with newspaper to catch drips.

Most of the components used in fiberglass work are irritants. In fact, both the resin and the hardener are caustic, flammable and toxic. Use them only in a well-ventilated area, away from any flame, and keep them away from skin and eyes. Always wear a respirator and goggles when you cut or sand fiberglass, since fine glass fibers will be released into the air.

A Two-sided Patch for a Plastic Laundry Tub

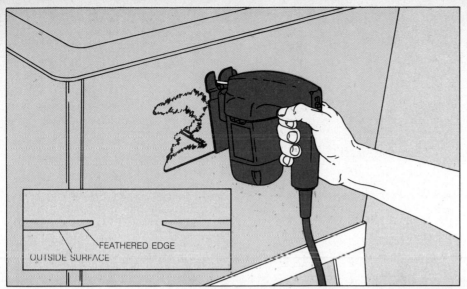

1 Cutting out the damage. After removing any facing or skirt from the tub, use a saber saw with a metal-cutting blade to remove damaged material. Make the hole a regular shape to simplify patching. Use a sanding block with 80-grit sandpaper to remove loose particles from the edges of the hole.

On the outside surface of the tub, feather a two-inch bevel around the hole (*inset*), using an electric drill with a sanding disk fitted with 80-grit paper. Cut a piece of cardboard slightly larger than the hole, cover it with plastic wrap, and then tape it to the inside of the tub so that the plastic wrap is facing the hole.

FEATHERED EDGE
OUTSIDE SURFACE

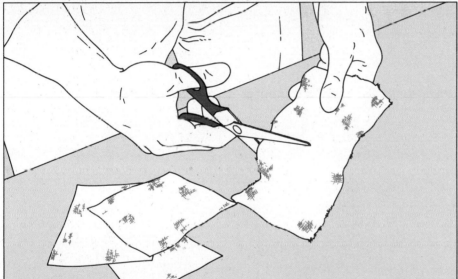

2 Cutting the patches. Cut a piece of fiberglass cloth large enough to cover the hole and the 2-inch feathered area, then cut a piece of fiberglass mat slightly smaller. Continue cutting alternate pieces of cloth and mat, each slightly smaller than the last, until the total thickness of the cut pieces equals the thickness of the tub wall. The final piece should extend about 1 inch beyond the hole on each side. Fray the ends of the glass fibers at the edge of each piece so that they can be pressed flat.

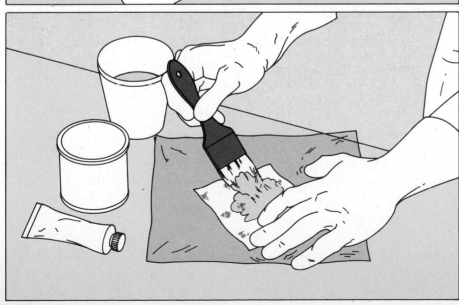

3 Saturating the fiberglass. Lay the largest piece of fiberglass cloth on a sheet of plastic wrap, mix the resin and hardener, and use a soft-bristled brush to saturate the cloth with resin. Work with a dabbing motion to avoid wrinkling the cloth. Center the largest piece of fiberglass mat on the cloth, and dab the mat with resin until it is saturated. Continue to add alternating layers of glass cloth and mat until you have built up the thickness that you need. Pick up the plastic wrap and transfer the patch to the outside of the hole, centering it over the opening and pressing it gently into place.

4 Pushing out air bubbles. With the plastic wrap still in place, use a plastic squeegee to press air bubbles—which appear as white spots—out of the patch, working from the center to the edges. Let the resin cure for several minutes, then carefully peel away the plastic wrap. Allow two to five hours for the patch to cure, then remove the cardboard backing.

Use an electric drill with a sanding disk to make a shallow depression in the patch on the inside surface of the tub, feathering the edge 2 inches beyond the original hole. Cut pieces of glass mat to fill the depression, ranging from the size of the original hole to 1 inch larger on each side. Cut two pieces of glass cloth the size of the feathered area around the hole.

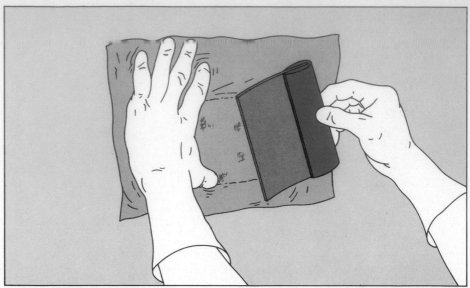

5 Applying the inside patch. After protecting the undamaged surface around the hole with paper and crepe masking tape, mix a new batch of resin and paint the sanded area with it, then apply the smallest piece of mat. Saturate the mat by dabbing on more resin with a brush. Repeat this process with successive layers of mat, then with the final pieces of cloth, using the brush to force out air bubbles in each layer before applying the next. The final layer should be slightly higher than the original surface of the tub. Lay a piece of plastic wrap over the entire patch, then press out any remaining air bubbles with a squeegee, working from the center outward.

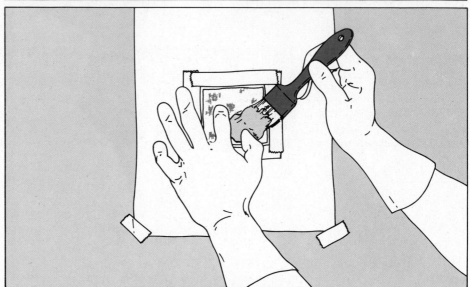

6 Trimming the patch. When the patch has partially hardened, carefully peel away the plastic wrap, then use a utility knife to cut away any loose edges or stray strands. Do not lift the patch or otherwise disturb the smoothed area.

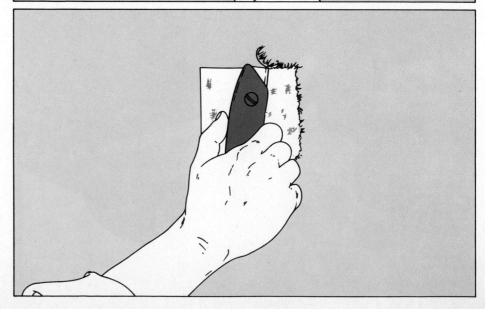

7 **Applying the gel coat.** Mix a gel coat (page 91) and use a soft-bristled brush to apply an even coat over the patch. Cover the gel coat with plastic wrap and use a squeegee to press out air bubbles, smoothing gently from the center toward the edges. Let the gel coat harden completely, then gently peel away the plastic wrap.

Lightly hand-sand the gel coat, using a fine-grit wet-or-dry paper covered with fiberglass rubbing compound. Brush on a second gel coat in the same manner as the first, allow it to cure, then polish it with a buffing wheel and rubbing compound (page 106, Step 6).

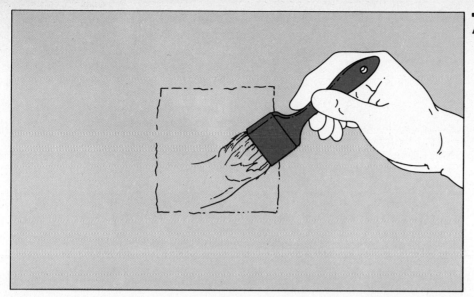

A Supported Patch for a Hard-to-Reach Place

Attaching backing from the front. To secure a fiberglass patch in a hole accessible from only one side, you will need to glue a piece of cardboard to the inside surface. First use a bastard-cut mill file to chamfer the edges of the hole on both sides, so that the hole is bordered by sharp edges. Next cut a piece of cardboard to overlap the chamfered area by ½ inch on all sides. Use cardboard flexible enough to bend through the hole but still retain its shape. Cut a hole through the center of the cardboard and push a thin bolt through, then tie a piece of string to the threads of the bolt. Apply a thin bead of quick-drying contact cement around the edge of the cardboard, on the same side as the string. Bend the cardboard slightly and push it through the hole (top left).

Pull on the string, then use the bolt to center the cardboard behind the hole. Pull the cardboard against the inside surface so that the cement is firmly in contact with that surface (center left). Slide a wood slat with a hole in it over the bolt, and tighten a nut on the bolt; these pieces will hold the cardboard in place while the adhesive is drying (bottom left).

Remove the nut and the wood slat; push the bolt through the cardboard or cut it off with a bolt cutter. Apply a layer of resin-soaked glass cloth inside the hole, tucking the edges into the space between the cardboard and the chamfered edge of the wall. Allow the resin to cure, then complete the patch as shown in Steps 5-7.

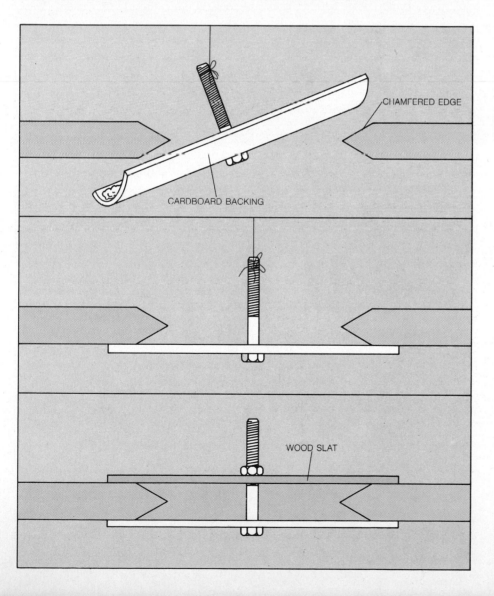

CHAMFERED EDGE

CARDBOARD BACKING

WOOD SLAT

A New Generation of Buffered Sealants

Polymers, the stuff of plastic resins, have infiltrated the manufacture of such age-old sealing agents as mortars, grouts and caulks—in some cases a polymer fortifies existing formulas, and in others it replaces the older material altogether. Most modern caulks are made completely of plastics; mortars and grouts are still largely cement, sand and water. But modified mortars and grouts supplement the standard versions and are useful where greater strength is needed.

The toughest of the fortified mortars is one that has an epoxy added—it is about three times stronger than conventional mortar. It is also more resistant to the elements, and it adheres to wood, a property that permits stone flooring to be laid over a wooden subfloor rather than over a concrete slab. For patching concrete, epoxy-fortified mortar is superior to conventional mortar.

Epoxy-fortified mortar has three ingredients—a liquid resin, a liquid hardener and a cement-sand powder; all are available from building-supply and hardware stores. The two epoxy components are mixed first, then the mortar powder is added, usually in a ratio of about 1 quart of epoxy mix to 13 pounds of powder. The epoxy components come in 1-pint to 5-gallon containers.

Working with epoxy mortars requires the same tools as for regular mortar—a mixing bucket, a stirrer and a notched trowel—with the addition of rubber gloves to spare your hands possible irritation from the epoxy. The working methods are also the same, except that epoxy mortars set up faster than regular mortars. Mixed, they have a pot life of about three hours; spread, the mortar must be topped by brick or stone in about 20 minutes, or a skin will form across the top, rendering further cementing impossible.

Latex, another plastic resin added to mortar, yields a product about twice as strong as conventional mortar and midway in cost between epoxy mortar and the conventional type. It is noted for its resistance to moisture and to cracking caused by vibration. But unlike epoxy mortar, it does not adhere to wood. Sold in two parts—a liquid resin and a cement-sand powder—latex mortar is mixed and applied in the same way as epoxy mortar.

Latex additives are also used to fortify grout, the cement mixture troweled between tiles or stone, giving this product advantages that latex gives to mortar: greater strength and resistance to water. Translated into practical terms, this means fewer repairs and easier cleaning.

Commonly available in tile stores, although at a premium price, latex grout comes in two parts— liquid latex and cement powder—to be mixed in the proportion of roughly 26 ounces of latex to 10 pounds of powder. No more difficult to use than regular grout and calling for no extra tools, the mixture—like other plastic sealants—must be applied faster than conventional grout.

Plastic caulks, like their predecessors, the linseed-oil-base colloids, are available in tube and rope form, as well as in cans and caulking-gun cartridges. Applied to outside surfaces, especially places where dissimilar materials meet, these modern caulks do a better job of keeping out weather and insects.

Of the many caulks on the market, the most common (chart, opposite) are acrylic latex, solvent acrylic, butyl, polysulfide, polyurethane and silicone. The two acrylic caulks are for indoor and outdoor work respectively. The butyl will last longer than these outdoors, and more durable still are the last three, called elastomers for their rubber-like properties. Unfortunately, some of these elastomers will shed paints that cannot flex readily under temperature changes, but to make painting unnecessary, they come ready-made in common colors.

Working with Resin-fortified Materials

Using resin-impregnated materials. Clean the work area and assemble the tools and materials you need. Mix the mortar or grout according to the manufacturer's instructions (here, the mortar is an epoxy type), and trowel out enough to last for about 10 minutes of work. For mortar, spread the mixture with a notched trowel held at a 45° angle (above, left). For grout, spread the mixture with a rubber float held at a 45° angle, forcing the grout into the spaces between the tiles (above, right). Use a finger to press grout into corners. Let the grout dry for 5 to 10 minutes, then wipe the tile with a clean, wet sponge. Rinse the sponge often in a bucket of water, but do not pour the grout-filled water in a sink or bathtub; the grout may clog the drain. Let the tile surface dry until a haze appears, then buff the tile with a crumpled piece of nylon.

Common Caulks and How to Use Them

	Durability	Adhesion	Tack-free	Solvent	Comments
Acrylic Latex	2 to 10 years	Good	¼ to ½ hour	Water	Not recommended for outdoor use or on metals
Solvent acrylic	2 to 10 years	Good	¼ to ½ hour	Toluene	Satisfactory on most surfaces; good for around doors and windows
Butyl	7 to 10 years	Excellent	½ to 1½ hours	Naphtha, paint thinner	For use mainly on masonry; high moisture resistance; shrinks slightly as it cures; principal color gray
Polysulfide	20 years	Excellent	24 to 72 hours	Toluene, TCE, MEK	Requires recommended primer on masonry; not recommended for use on glass; available in aluminum color for use on storm windows
Polyurethane	20 years	Excellent	24 hours	Toluene, paint thinner, MEK	Relatively easy to apply; requires recommended primer on metal
Silicone	20 years and up	Excellent	1 hour	Naphtha, toluene, xylol, paint thinner	Excellent around bathtubs and outdoors on glass and metal; requires primer on masonry

Ranking caulks. The chart above shows common kinds of plastic caulks along with their main distinguishing features. Listed horizontally, the caulks are rated for durability and adhesion, and for the length of time they take to lose their initial stickiness, which indicates how long they stay workable. Also given with each listing is the solvent needed to smooth or clean the caulk; the final column shows special factors governing the use of a given caulk. All the caulks are available in colors, all but the latex caulks are suitable for outdoor work and, in general, the longer lasting the caulk, the higher its price. The solvents abbreviated are trichloroethylene (TCE) and methyl ethyl ketone (MEK).

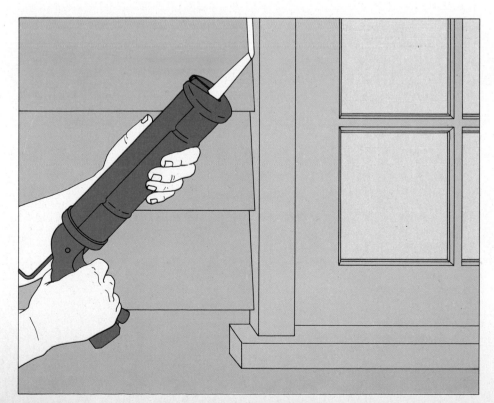

A Simplified Seal Done with a Caulking Gun

Applying the caulk the right way. Before beginning the caulking operation, check the condition of the surface to be caulked; except when you are using latex caulk, which can be applied on a damp surface, the area should be clean and dry. Also check the cartridge label for special instructions covering such situations as caulking in cold weather or over unprimed surfaces. Then prepare the cartridge: Cut off the nozzle at a 45° angle and puncture the inner seal by pushing a long nail into the nozzle. Hold the cartridge against the work at a 45° angle and pull it toward you, pumping the trigger. Make the bead of caulk overlap both edges of the crack. Smooth the bead with a solvent-soaked cloth.

Plastic-Base Coatings: Paint for Any Purpose

Except in historic restorations, where paints based on linseed or tung oil are still used, most modern coating materials are made of plastics; alkyd- and latex-base paints are practically universal for exterior house painting, and latex is the standard choice for interior work. In addition, a great array of fluid resins, from synthetic rubbers to two-part epoxy coatings, make it possible to coat not just wood or wallboard or masonry, but virtually any surface in or on your house.

You can, for example, give a factory-like finish to plastic, metal or ceramic fixtures—from kitchen-range hoods to bathroom sinks—with polyurethane, vinyl or epoxy coatings that will dry to an ultrahard sheen or to a rubbery, grease-resistant hide. With virtually no preliminary preparation, you can convert slabs of rough wood—such as barn boards or slices of tree trunk—into attractive table-tops by coating them with a thick, tough, crystal-clear layer of a specially formulated epoxy resin. And tool handles dipped in a rubber-base compound instantly acquire slip-proof, cushioned grips.

Outside the house, weathered shingle roofs can be covered with flexible alkyd coatings that expand and contract with temperature changes. Clear silicone protects brick walls from the ravages of moisture. Oil-resistant, fast-drying acrylic toppings sheathe driveways; skidproof compounds based on synthetic rubbers can protect the surfaces of walkways and steps and make them safer to tread on when the day is rainy.

Of course, no single resin can offer every advantage of today's modern coatings. Yet so versatile are plastic resins that there is usually a selection of coatings for a given job. Making a choice from among the available products is the first crucial step in any painting or sealing job.

To guide you in this choice, the box below lists the major families of coatings and indicates the results that each one delivers. Consult the list to see which products seem likely to serve your needs; then, having narrowed your choice, consult the chart at right to see which of these are compatible with the surface that is to be covered. If the surface has previously been painted, you should also consult the chart on page 118 to find out which of your choices will be compatible with the old coating.

Finally, consider the cost of the materials that are still in the running. You may reject some high-performance coatings—especially those that are based on epoxy, vinyl or urethane resins—as being too expensive for a particular job, even though they might work somewhat better than the runner-up.

Pros and Cons of Frequently Used Finishes

Included in this guide are the eight major families of plastic-base coatings. The descriptions include only those properties offered by the product in its most common form. Many large paint stores also carry special formulations—such as polyurethane floor coatings with admixtures of nonskid chips.

□ ALKYDS. Available clear or in colors, alkyd-base paints cover easily and are economical. They adhere well to most surfaces except fresh concrete and plaster. Alkyds are durable under normal conditions but deteriorate with excessive exposure to detergents, solvents and corrosive chemicals.

□ EPOXIES. These coatings are produced by a chemical reaction between an epoxy resin and a hardening agent, mixed together before application. The mixture retains its working consistency for 15 minutes to four hours; excess cannot be saved. Epoxy paints, available in all colors, adhere excellently to wood, metal or masonry. They form tough, hard coatings that can withstand abrasion, solvents, detergents and corrosive chemicals. High price limits epoxy to jobs where high performance is a must.

□ LATEXES. Latex paints—which are also called acrylic latex—are economical and easy to apply; in addition, they are nonflammable. They resist fading and stand up to light abrasion and corrosion. Latex films breathe, so they resist blistering, but they require thorough surface preparation for good adhesion.

□ PHENOLICS. Phenolic resins are most often combined with natural oils to provide a versatile, moisture-proof base for clear varnishes and enamels. Such products dry fast and are especially durable in humid environments. They are expensive, however, and they are not available in white or light colors because the resins have a yellowish tint.

□ POLYURETHANES. Polyurethane coatings cure hard, glossy and tough, with excellent chemical and abrasion resistance. But polyurethanes are highly flammable until they dry, and they are expensive. Because they contain solvents that attack existing coats of alkyd or latex paint, they cannot be applied over those coatings. Clear polyurethane darkens with long exposure to sunlight.

□ SILICONES. Dilute solutions of silicone resin will shield unpainted masonry surfaces against moisture without changing their color or appearance. Higher concentrations may be combined with alkyds and aluminum or carbon pigments to produce stove paints that withstand temperatures up to 1,200° F.

□ SYNTHETIC RUBBER. Chlorinated polyisoprene, polystyrene butadiene, polyvinyl toluene and other resins—synthetic rubbers—all form the basis for many water-resistant coatings. Available in a moderate range of colors, such rubber-base paints seal and decorate swimming pools and masonry; they are a good choice for walls that are subject to condensation or frequent washing.

□ VINYLS. Resins derived from polyvinyl chloride and polyvinyl acetate produce flexible coatings with exceptional durability and resistance to water, corrosive chemicals and abrasives. Most commonly available in combination with alkyd resins, vinyl coatings are ideal for metal and masonry surfaces in marine use and where chemical resistance and good appearance are important.

Matching the Paint to the Stresses It Will Face

Material	Use	Environment	Resin Alkyd	Epoxy	Latex	Phenolic	Polyurethane	Silicone	Synthetic rubber	Vinyl
Wood	General purpose	Interior	●		●		●			
		Exterior	●		●					
		Exterior/marine	●				●			
	Heavy duty	Interior	●	●	●	●	●			
		Exterior	●			●	●			
		Exterior/marine	●				●			
	Floors	Interior	●				●			
		Interior/high abrasion	●				●	●		
Metal	General purpose	Interior	●							
		Interior/high humidity		●		●	●		●	
		Exterior	●			●				
	Heavy duty	Exterior/marine		●		●	●		●	●
		Exterior/corrosive		●			●			●
	Floors and stairways	Interior	●	●		●	●			
Plaster and wallboard	General purpose	Interior	●		●					
		Interior/high humidity	●				●		●	
Masonry	General purpose	Interior	●		●		●			
		Interior/high humidity		●			●		●	
		Exterior	●		●			●	●	
	Heavy duty	Exterior/marine		●				●	●	
		Exterior/corrosive	●	●						●
	Floors	Interior	●				●			
		Interior/high humidity	●				●		●	
		Interior/heavy traffic	●				●			
Glass	General purpose	Interior	●	●	●					
Plastic	General purpose	Interior	●	●						

Choosing a paint. Surfaces vary not only in material but in the kind and amount of punishment they must take; the three columns on the left of this chart are in effect a job description of the project. "Heavy duty" describes a surface that must be frequently washed or is subject to abrasion. "Marine," "high humidity" and "corrosive" refer to environment—the first to coastal areas with salty air; the second to places subject to dampness, such as bathrooms; the third to urban locations, where the air may have high proportions of smoke and acidic gases. Find the line for your project, then read across to the coating column on the right. A dot under a coating means it is acceptable for the job.

Preparing a Compatible Surface for Paint

Plastic-base paints, no matter how superior their credentials, will perform well only on a surface that has been properly prepared. The surface must provide a strong base for the new coating. If it has weak spots or is rough or uneven, follow the steps on pages 105-115 to reinforce, smooth and fill it. After this stage, preparation varies with the type of surface and whether it is already coated.

If the surface has been previously painted or sealed, first make sure that the coating you are planning to apply is compatible with the old coating (chart, below, right). Once you know that the coatings are compatible, roughen any glossy areas of the old paint with sandpaper, steel wool or a commercial deglossing solution, obtainable at paint stores. Flaking or peeling paint must be taken off completely; hand scrapers, chemical paint removers and heat guns are good for this job. Large exterior metal and masonry surfaces, such as wrought-iron fencing and brick walls, can be cleaned with power tools—sandblasters for metal, waterblasting equipment for masonry, both available at tool-rental stores.

Surfaces to be painted for the first time must be scrupulously clean. Remove grease and oil from wood with mineral spirits. Seal knots, which may ooze sap, with a commercially available liquid compound, generally a blend of polyvinyl butyral and a phenolic resin.

Metal surfaces are best cleaned with acetone or lacquer thinner, or with more expensive commercial degreasing compounds. Less costly alkaline cleaning agents—strong detergents and trisodium phosphate—are also effective; however, these cleansers should be dissolved in hot water—150 to 200° F.

Steel and iron must also be stripped of any rust. Treat light rust with phosphoric acid in gel or solution form, available at hardware and paint stores. Not only do they clean away rust, they also etch the metal surface to increase paint adhesion. For serious rust, chip or scrape away the loose scales with a wire-brush attachment on an electric drill or with a wire brush and a hand-held scraper.

Bare concrete and masonry-block surfaces present two special problems. Porous types are difficult to clean—dirt and grime settle deep into crevices and can cause paint to blister or peel. Smooth, glazed masonry, on the other hand, provides a difficult surface for any new coating to penetrate.

To clean masonry, a stiff brush and strong detergent are effective, but a high-pressure spray-cleaning device—obtainable at tool-rental agencies—will save considerable time; it can pump a jet of water at 600 pounds per square inch to blast grime off a walkway or a wall. Once clean, bare masonry should be treated with muriatic acid to etch the smooth glazed surfaces—painters call it "raising a tooth"—for the new coating to grip. Muriatic acid also will dissolve efflorescence, a powdery alkaline substance that sometimes crystallizes on the surface of damp brick or cinder block. Afterward, rinse the surface with water to remove the acid. All masonry should be allowed to weather for as long as possible before painting; freshly poured concrete needs 60 to 90 days to release its alkalinity.

Plastic surfaces, though smooth, usually do not need to be roughened for good adhesion. Static electricity, however, can sometimes cause paint to crawl, or alligator, on plastic surfaces—especially acrylic sheets—and must be removed with an alcohol wash (opposite, bottom). If the object being painted is portable, an electrically grounded metal screen placed underneath the plastic will help to keep it static-free during painting.

Surface preparation can be tedious and messy. When working with caustic cleaning agents, wear a hat, rubber gloves and old clothes. Wear a dust mask for fine-sanding projects, and don goggles and a charcoal-cartridge respirator before using muriatic acid indoors. Ventilate your work area well; also, before you start, clear away or cover anything that could be damaged by spills.

Making a Chemical Match with Paint over Paint

Existing surface	New Coating	Alkyd	Epoxy	Latex	Phenolic	Polyurethane	Synthetic rubber	Vinyl
Alkyd		●	○	●	●	○	○	○
Bituminous (roof coating)		○	○	●	○	○	●	●
Cement paint		○	●	●	○	●	●	●
Epoxy		○	●	○	○	●	○	●
Latex		●	●	●	●	●	●	●
Oil		●	○	●	●	○	○	○
Phenolic		●	○	●	●	○	○	○
Polyurethane		○	○	○	○	●	○	○
Silicone		○	○	○	○	○	○	○
Synthetic rubber		●	○	●	●	○	●	●
Vinyl		●	●	●	○	○	●	●

Recoating an old surface. To determine whether a new paint can safely be applied over an existing paint, read down the first column to find the current surface paint. Below each possible new coating listed across the top, a solid dot indicates compatibility. An open dot indicates that the pairing is not recommended. If you do not know what kind of coating is currently on the surface, experiment with a small amount of the new coating in an inconspicuous place.

An Acid Scrub for Stains

Cleaning masonry. Use a wire brush to remove loose efflorescence. Then, wearing rubber gloves and protective clothing, paint on a solution of one part muriatic acid to nine parts water. To avoid dangerous foaming and splattering, always pour the acid into the water, never the water into the acid. Apply the solution to the masonry with a natural-bristled scrub brush, using plastic dropcloths, if necessary, to keep the solution off anything except the masonry. When the acid has stopped bubbling, rinse the surface thoroughly with water.

Draining Static Electricity

Ridding sheet plastic of static electricity. Loop one end of a bare copper wire through a piece of metal screening and attach the other end to an electric ground, such as a metal cold-water pipe or the small screw that secures a cover over an electric receptacle. Place the screen on the worktable, under the plastic sheet, and wipe the plastic with a mixture of 1 part isopropyl alcohol to 9 parts water. Although any soft cloth can be used to wipe on the alcohol, a well-wrung chamois will result in the least spotting.

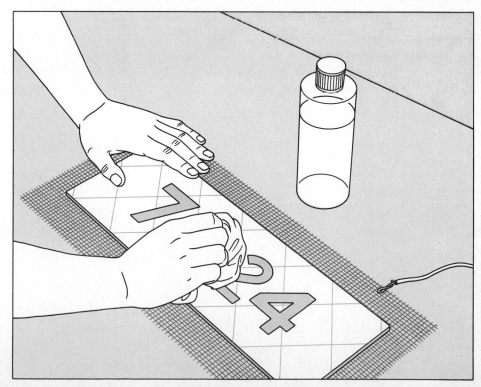

The Fine Points of Application

Plastic-base coatings vary widely in both consistency and composition. Some are watery-thin for fast spreading; others are soupy for a textured finish. Some are made with water for easy cleanup, others with volatile solvents for quick drying. The variety of forms in which plastic-base coatings are produced has spawned an equally broad range of techniques for applying them to surfaces.

Brushes, of course, are the standard tool, but some coatings require natural-bristled brushes, whereas synthetic bristles work better for others. For general-purpose coatings, such as alkyds and phenolics, use a natural-bristled brush of good quality. Latexes demand nylon or polyester brushes, which do not get soggy in water. Fast-curing, high-gloss epoxies and polyurethanes—often applied to metal or very smooth wood—are spread best with disposable sponge brushes. And there is a special nylon-bristled rough-surface brush that simplifies the job of working both latex and alkyd coatings into the cracks and crevices of stucco and brickwork.

The correct way to use the paintbrush also varies with the type of coating you have chosen to apply. Alkyds and latexes should be brushed thoroughly—first back and forth, then crosswise. Synthetic-rubber coatings, in contrast, should be brushed on with two or three strokes—just enough brushwork to spread them evenly. Clear phenolics and polyurethane coatings ought to be flowed onto the surface, as slowly and deliberately as possible.

For large surfaces, rollers spread coating films more smoothly and rapidly than brushes. A fully loaded 9-inch roller can hold enough paint to cover about 6 square feet of surface, compared with 1 square foot for a 3-inch brush. Like brushes, however, rollers have to be matched to the coating being applied. Natural lamb's wool, for example, is unaffected by solvents, so it is suited to volatile epoxies and polyurethanes. On the other hand, the fibers of lamb's wool are likely to mat, and even fall out, after contact with alkaline latex paints. Use the charts below and opposite to choose the right roller for your work.

Specialized applicators for hard-to-cover surfaces include paint pads, which are helpful for areas too confined for the use of rollers, and lamb's-wool painting mittens, which are used for wiping paint onto thin posts and rails. And some coatings are formulated to be applied without any tools at all; they are simply poured onto a surface. Or, in some cases, the object to be coated is just dipped and allowed to drain.

Fitting the Roller Fabric to the Paint

Coating	Smooth	Surface / Textured	Rough
Alkyd	Lamb's wool	Acrylic	Acrylic
Epoxy	Mohair	Lamb's wool	Lamb's wool
Latex	Polyester	Polyester	Polyester
Phenolic	Mohair	Acrylic or polyester	Acrylic or polyester
Polyurethane	Mohair	Lamb's wool	Lamb's wool
Silicone	Mohair	Lamb's wool	Lamb's wool

Choosing the roller fabric. To match a roller to a painting job, consider both the paint and the surface. The fabric of the roller cover must spread a paint evenly without reacting to its resin. It also must travel over a surface of a particular texture without matting or falling apart. Use this chart to determine the recommended roller fabric, then consult the chart on the following page to find the proper nap length.

Matching the Roller Nap to the Surface

| | Surface | | |
Roller fabric	Smooth	Textured	Rough
Acrylic	¼″ to ⅜″	⅜″ to ¾″	1″ to 1¼″
Lamb's wool	¼″ to ⅜″	¼″ to ¾″	1″ to 1¼″
Mohair	³⁄₁₆″ to ¼″		
Polyester	¼″ to ⅜″	½″	

Choosing the nap length. Once you have matched the fabric to the type of paint you are using, consult this chart to match the length of the nap to the surface being covered. In general, longer naps hold more paint, which makes them best suited for rough surfaces. For use on extremely smooth surfaces, such as wood paneling, special mohair rollers are available with nap fibers as short as ³⁄₁₆ inch.

Special Tools for Special Situations

Coating rough surfaces. To work paint into the cracks and crevices of a coarse surface, use a rough-surface brush with nylon bristles that are not only thicker than ordinary, but split, or flagged, at the ends to carry more paint. Dip the brush squarely into a pan or roller tray containing about ½ inch of paint. Work it over the surface with an up, down and sideways spreading motion. For particularly hard-to-reach areas, such as mortar joints or the undercut divisions between shingles or shakes, tilt back the brush and use the narrow row of bristles on one edge.

Applying an ultraglossy coating. To keep brush marks from showing on a mirror-smooth surface, use a sponge brush rather than a bristle brush. Submerge about half the sponge in the paint, then paint with short, steady strokes, taking care that you allow only the beveled tip to come in contact with the work.

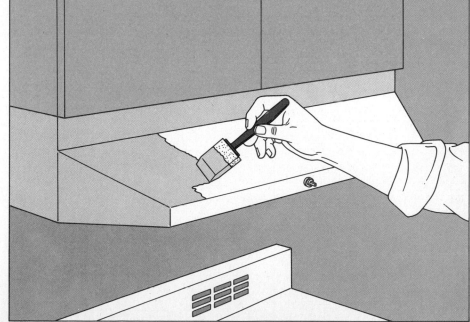

The Pour-on Polymer Coatings

1 Combining the components. Mix equal parts of resin and hardener in a disposable calibrated paper cup. Stir the batch thoroughly but gently with a clean wooden stirring stick, taking care not to introduce air bubbles. After two minutes of mixing, pour the thick coating over the surface to be covered—here, a butcher-block tabletop—working in a spiral from the outside in.

Dipping to Sheathe a Handle

Sheathing an object in plastic. Suspend the object to be coated on a length of twine or wire, and slowly lower it into a can of rubber-base vinyl plastic coating compound. Withdraw the object slowly, about one inch every five seconds, and hang it to dry. Immediately seal the open can; escaping vapors could soften the sheathing, causing it to drip or sag. After about 20 minutes, dip the object a second time. Then allow the sheathing to dry thoroughly to a tough, pliable finish; this will take about four hours.

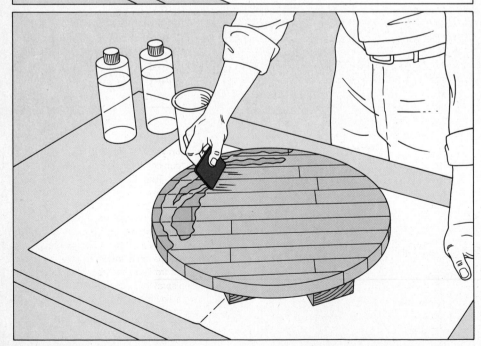

2 Smoothing the coating. Using a plastic-covered playing card as a spreader, gently distribute the coating evenly over the surface. Allow some of the coating to run over the edge, as in icing a cake; smooth this edge coating with a disposable sponge brush. Prick any large bubbles with a toothpick, then snare stray bits of lint with tweezers. While the coating is soft, breathe on the surface with your mouth open—the carbon dioxide will break tiny surface bubbles.

When the coating is completely dry, in two or three days, add additional layers if desired, until the coating is as much as ¼ inch thick.

Spraying the Professional Way

The versatility and adaptability of plastic-base paints and sealers suits them to the fastest of all application techniques: spray-painting. With a portable spray gun, it is possible to apply a uniform, mirror-smooth coating more quickly than with a brush or a roller, or by any other manual method.

Virtually any large surface, from fine wood to coarse cinder block, can be spray-painted. Even small surfaces and trim can be sprayed with careful masking, although for very small areas the time needed for masking may exceed the time saved by spraying.

Equipment for spray painting ranges in complexity from disposable, albeit expensive, aerosol cans, which deliver a fine coating mist for small finishing projects, to high-capacity compressor-driven spray guns that can be rented from tool-supply stores. Between these extremes lies a tool that can perform almost all of the jobs handled by conventional compressor-powered units, and with a real savings in paint: the lightweight airless sprayer. Conventional sprayers use compressed air to create a misty mixture of paint and air. The airless guns employ a small but powerful electric pump to propel only droplets of paint through their nozzles, yielding more precise application and far less overspray than the compressor-driven sprayers.

Before painting with any airless sprayer, familiarize yourself thoroughly with its operation. Because models differ from manufacturer to manufacturer, read the manual supplied with your sprayer—and note the safety precautions in the box at right. Then, if you have never spray-painted before, fill the spray gun with water and get the feel of using it by spraying newspapers hung vertically.

Although virtually any plastic-base coating can be sprayed, it generally must be diluted before use (page 124). Most sprayers require paint about 25 per cent thinner than brushing consistency. The paint must also be free of impurities. If you notice any floating skin or dirt, filter the paint through twice-folded cheesecloth. Remember, too, that careful surface preparation is just as important for spraying as for any other method of application (page 118).

In addition, spray painting calls for extra caution in preparing the surfaces that you do not wish to paint. Because atomized paint can get picked up by air currents and drift a considerable distance—especially outdoors—protect the nearby surfaces with dropcloths and newspapers held down with masking tape.

Finally, whenever you finish using a spray gun, clean its mechanism by flushing it with a solvent compatible with the paint that was sprayed.

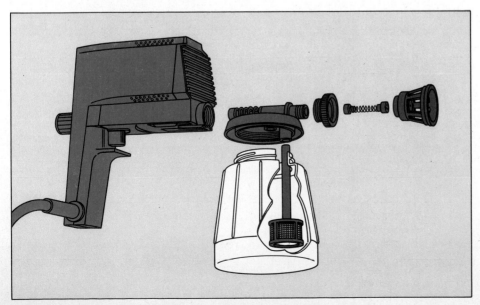

The airless sprayer. The motive force for an airless spray gun comes from a powerful electro-magnetic motor. Turned on and off by a trigger switch—and regulated in velocity by a control knob on the rear of its insulated housing—the motor drives a spring-loaded piston back and forth. The moving piston sucks paint from the container, through a mesh filter and suction tube, to the cylindrical paint chamber, and then forces the paint from the chamber, past the spring-loaded spray valve and out through the nozzle, whose tiny orifice directs minute droplets of paint toward the surface being coated.

For large painting jobs, the small paint container may be removed and an extended exhaust suction tube (not shown) fitted directly to the nipples on the bottom of the paint chamber. The long tube, fitted with a clip and a filter, is then immersed directly in a large bucket of paint.

Readying paint and sprayer. Dip a viscosity measuring cup, which commonly comes with an airless sprayer, into the paint. Lift the cup, and time how long it takes for the paint to drain out through the hole in the bottom; compare this time with the time the sprayer manufacturer recommends. If necessary, mix in a small amount of the proper solvent, then test the consistency again. Continue thinning and testing until the paint runs out the hole within the time recommended for your sprayer.

Filter the paint through cheesecloth if necessary, and pour it into the paint container of the sprayer, filling the container. Aim the gun at newspapers taped to a wall, and adjust the control knob until the paint sprays out in a uniform wedge-shaped pattern. Spattering paint indicates that the velocity is set too low; paint running down the paper indicates too high a setting.

A proper indoor spraying setup. To guarantee a safe and neat job, small surfaces you do not wish to coat should be covered with strips of masking tape. Shield larger areas with dropcloths. Wear a charcoal-cartridge respirator when spraying, and install a fan in the nearest window to draw away fumes. If you are far from a window, carry fumes to it by rigging plastic dryer duct from the fan to the work area. Keep a fire extinguisher—either a carbon-dioxide or a dry-chemical one—handy in case of fire.

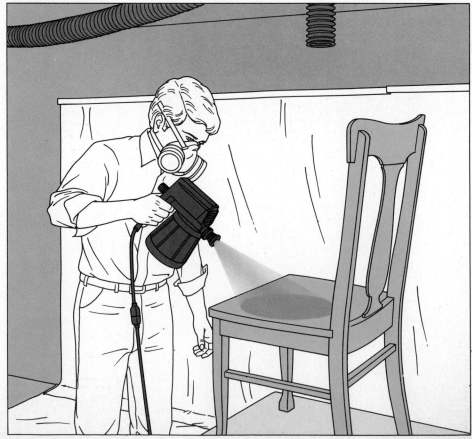

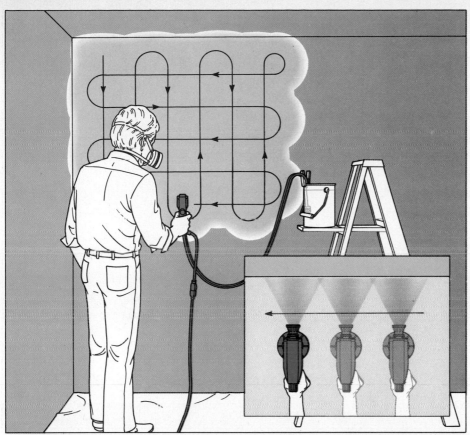

Spraying walls. For large jobs, supply paint to the sprayer through an extension tube rigged to a large paint container. Hold the gun horizontal to the wall surface and, while spraying, move your entire arm up and down or left and right, keeping the nozzle about 8 inches from the wall (*inset*); do not swing your arm in an arc, which would vary the distance of the nozzle from the wall, bringing it closer at the center of each sweep, farther at the ends. Large, smooth walls are best covered with a crosshatch pattern, the paint applied first with vertical passes, then horizontal ones. For walls that are divided horizontally by moldings or clapboards, use overlapping horizontal passes only.

Coating horizontal surfaces. To paint ceilings or floors, replace the regular nozzle with a flexible extension nozzle. The extension—available for most airless sprayers—is made of soft copper tubing that you can bend with your fingers, thus aiming the spray up or down at horizontal surfaces without making it necessary for you to tip the gun or aim it directly overhead.

125

Picture Credits

The sources for the illustrations in this book are shown below. The drawings were created by Jack Arthur, Roger Essley, Charles Forsythe, William J. Hennessy Jr., John Jones, Dick Lee, John Martinez and Joan McGurren. Credits for the illustrations are separated from left to right by semicolons and from top to bottom by dashes.

Cover: Fil Hunter. 6: Fil Hunter. 11-19: John Massey. 21-31: Frederic F. Bigio from B-C Graphics. 33-36: Walter Hilmers Jr. from HJ Commercial Art. 38-41: Arezou Katoozian. 42: Fil Hunter. 44-47: Eduino J. Pereira from Arts and Words. 49-53: William J. Hennessey Jr. 54-67: John Massey. 68: Fil Hunter. 71-77: Arezou Katoozian. 78-83: Walter Hilmers Jr. from HJ Commercial Art. 85-89: Eduino J. Pereira from Arts and Words. 92-95: Frederic F. Bigio from B-C Graphics. 97: John Martinez. 99-101: Elsie J. Hennig. 102: Fil Hunter. 105-109: Frederic F. Bigio from B-C Graphics. 111-113: Stephen A. Turner. 114, 115: Elsie J. Hennig. 119-125: Frederic F. Bigio from B-C Graphics.

Acknowledgments

The index/glossary for this book was prepared by Louise Hedberg. The editors wish to thank the following: Edna F. Ashley, Arts and Crafts Director, Ft. McNair, Washington, D.C.; Charles Bender, Design Engineering, Hyattsville, Md.; Ray Bender, Design Engineering, Hyattsville, Md.; Louis Block, Graco Inc., Minneapolis, Minn.; Frank Devlin, H. B. Fuller, Construction and Consumer Products Division, Palatine, Ill.; Dow Chemical Co., Freeport, Tex.; Stephen Draper, Greenbelt, Md.; Elizabeth Durfee Hengen, Historic Preservation Consultant, Winchester, Mass.; Charles Hughes, Fairfax, Va.; Robert Kirkpatrick, The Gibson-Holmas Co., Twinsburg, Ohio; Jack Klarquist, Shell Oil Company, Houston, Tex.; John Lawrence, The Society of the Plastics Industry, New York, N.Y.; Susan Layden, The Flecto Company Inc., Oakland, Calif.; William Mazzerela, The Rohm and Haas Co., Philadelphia, Penn.; Maureen Migdalen, The Society of the Plastics Industry, New York, N.Y.; Don Morris, Fiberglass Engineering Co., Manassas, Va.; Jules R. Panek, Yardley, Penn.; Read Plastics, Rockville, Md.; The Rohm and Haas Company, Philadelphia, Penn.; Larry Russell, The Rohm and Haas Company, Philadelphia, Penn.; Robert Sellers, Smooth-On Inc., Gillette, N.J.; Shell Oil Co., Houston, Tex.; Specialty Products Company, Jersey City, N.J.; Strux Corporation, Lindenhurst, N.Y.; Jody Swisher, Wagner Spray Tech Corporation, Minneapolis, Minn.; Thermoset Plastics, Indianapolis, Ind.; Brian Tyack, Perma-Brite/TAP, Alexandria, Va.; Will Wharton, Dow Chemical Co., Technical Products Research Division, Freeport, Tex. The editors would also like to express their appreciation to William Doyle, David Shapiro and Wendy Shay, writers, for their help with the preparation of this volume.

Index/Glossary

Included in this index are definitions of many of the technical terms used in this book. Page references in italics indicate an illustration of the subject mentioned.